W9-CII-587

MATHEMATICAL MOUNTAINTOPS

MATHEMATICAL MOUNTAINTOPS

The Five Most Famous Problems of All Time

John L. Casti

2001

OXFORD
UNIVERSITY PRESS

Oxford New York
Athens Auckland Bangkok Bogotá Buenos Aires Calcutta
Cape Town Chennai Dar es Salaam Delhi Florence Hong Kong Istanbul
Karachi Kuala Lumpur Madrid Melbourne Mexico City Mumbai
Nairobi Paris São Paulo Shanghai Singapore Taipei Tokyo Toronto Warsaw

and associated companies in

Berlin Ibadan

Published by Oxford University Press, Inc.
198 Madison Avenue, New York, New York 10016

Library of Congress Cataloging-in-Publication Data is available
ISBN: 0-19-514171-7

Book design by Susan Day

1 3 5 7 9 8 6 4 2

Printed in the United States of America
on acid-free paper

CONTENTS

PROLOGUE

THE ART OF THE PROBLEM

THE MOST BEAUTIFUL?

In his *Metaphysics,* Aristotle says, "Those who assert that the mathematical sciences say nothing of the beautiful are in error. The chief forms of beauty are order, commensurability and precision." An even stronger assertion of the artistic character of mathematical enterprise came from John von Neumann, who remarked in his essay, *The Mathematician,* "I think it is correct to say that [the mathematician's] criteria of selection, and also those of success, are mainly aesthetical." Such quotes fill the literature of mathematicians talking about the emotional side of their work, and bear strong testament to how much closer the mathematician's mind is to that of the poet or artist than to that of the engineer. Contrary to the belief of even the educated layperson, mathematics is not about numbers and formulas but about patterns and structures. The patterns the mathematician discovers are then enshrined in what might be termed "The Book," the theorems constituting the currency of the mathematical profession.

In a very illuminating survey taken in 1990, popular mathematics writer David Wells asked the mathematical community to pick the most beautiful theorems (patterns) of all time. Wells chose 24 famous theorems ranging from the infinity of the prime numbers to the seemingly odd fact that every natural number greater than 77 is the sum of integers, the sum of whose reciprocals is 1. The full list of Wells's "beauty contestants" can be found in the paper cited in the Suggested Readings. But to get a feel for the variety of mathematical pulchritude on display, here are just a few of the candidates:

- $e^{\pi i} + 1 = 0$
- There is no rational number whose square is 2
- Every planar map can be colored using at most 4 colors
- At any party, there is a pair of people who have the same number of friends present
- $1 + \frac{1}{2^2} + \frac{1}{3^2} + \frac{1}{4^2} + \cdots = \pi^2/6$
- A continuous transformation of the points of a closed circle into itself has a fixed point

Wells received 68 usable responses to his questionnaire, which asked the respondents to mark each candidate on a standard scale of 0 to 10 for their beauty. The winner with an average score of 7.7 was Euler's formula, $e^{\pi i} + 1 = 0$, which links the five most important constants in mathematics in a single, compact, unexpected pattern.

In his comments on the replies, Wells notes several themes. One of the most important is that no criteria for beauty turn up more often than simplicity and brevity. Euler's formula certainly has these qualities in spades. Another common characteristic of beautiful theorems is that they are surprising. Surprise and novelty are expected to provoke emotion. But what happens when the novelty wears off? Surprise is also associated with mystery. Einstein once remarked, "The most beautiful thing we can experience is the mysterious. It is the source of all true art and science." It's a mystery itself as to how mathematicians distinguish between merely "pretty" and "beautiful."

This statistically highly insignificant survey did not lead to much by way of general conclusions, other than the negative one that the idea that mathematicians largely agree in their aesthetic judgments is at best grossly oversimplified. Wells concluded his analysis of the replies by saying that mathematics can really only be understood in the context of all human life. In particular, philosophies of mathematics that ignore beauty will be inherently defective and incapable of effectively interpreting the activities of mathematicians.

In this book we are concerned with exactly this theme: the activities of mathematicians. Specifically, our concern will be with what mathematicians do when they scale the tallest peaks in the mathematical mountain ranges, the hitherto unclimbed peaks that have challenged generations of mathematicians to scale them. Every one of the theorems on Wells's list started its life as one of these peaks. Those peaks have now all been scaled. The story of five such peaks forms the content of this volume. A companion volume will tell the tale of thus far unsuccessful attempts to scale five other peaks in the same range. As the stories told here outline the variety of equipment—mental and physical—needed to climb to such heights, we will not elaborate upon them now. Rather, it's of interest to have a deeper look at the peaks themselves. Where do "good" peaks to climb come from? How do certain peaks become "famous" challenges that all mathematicians try to climb at one point or another in their career? How do we know when we've actually arrived at the top of one of the peaks? These questions—and their answers—constitute an important part of the mathematical enterprise and deserve careful consideration.

PROBLEMS FOR ALL OCCASIONS

What does mathematics *really* consist of? Axioms (like the parallel postulate of Euclidean geometry)? Proofs (such as Gödel's proof of the undecidability of arithmetic)? Concepts (such as sets and numbers)? Definitions (such as Hausdorff's definition of dimension)? Theories (like number theory)? Formulas (such as the Cauchy integral formula of complex analysis)? Methods (like the method of steepest descent)?

All of these are essential to the mathematical enterprise and mathematics could certainly not exist without them. But none of them is at the very heart of the subject. The real raison d'etre for the mathematician's existence is simply to solve problems. So what mathematics really consists of is problems and solutions. And it is the "good" problems, the ones that challenge the greatest minds for decades, if not centuries, that eventually become enshrined as mathematical mountaintops.

There is a long tradition in mathematics that when a problem gets promoted to a mountaintop, someone directs attention to it by establishing a prize for its solution. While the monetary reward for climbing the peak can be substantial, the real reward for the mathematician is more likely to be a kind of one-upmanship on his colleagues, coupled with a deep sense of personal satisfaction at having seen the view from the top of the peak before anyone else. After all, people climb mountains for many reasons. But one of the most common is simply that they're there!

For a famous example of a "bounty" offered for the solution to a mathematical problem, think about the motion of the planets as they move about the Sun under the influence of their mutual gravitational attractions. It's a pretty natural question to ask if there might be some time in the future, however distant, at which two (or more) planets might collide? Or perhaps a time at which one planet acquires such a great velocity that it flies off out of the system? Translating this problem in physics to mathematical terms by assuming things like N perfectly spherical planets whose mass is concentrated at a point and motion in the absence of friction, dynamical system theorists of the latter part of the nineteenth century christened this the *N-Body Problem.*

In 1885, Swedish mathematician Gösta Mittag-Leffler convinced King Oscar II to offer money and a gold medal to the first person to find a general solution to the problem. By that was meant a representation of the coordinates of each of the N points in some known function of time (technically, the contest rules required this function to be expressed as a convergent power series, but that condition is really irrelevant to our discussion here). As it turned out, no one could produce a convincing solution and the prize was finally awarded to the French polymath, Henri Poincaré, for a partial solution—but one that led to the establishment of much of the modern theory of dynamical systems.

Here we see a common thread running through good problems: they tend to lead to the development of entirely new fields of mathematics. Poincaré's efforts on the N-Body Problem, even though they failed to actually solve the problem, laid the foundations of the modern qualitative theory of differential equations, which in turn has evolved into what is today termed *dynamical system theory.* Amusingly, modern work in dynamical system theory by Don Saari and Jeff Xia at Northwestern University in the U.S.A. has all but closed out the N-Body Problem by producing examples of specific systems for which particles do escape off to infinity.

It's worth noting in passing that shortly before the announcement of the King Oscar II Prize, another prize was offered by Paul Wolfskehl in Germany for the solution of the (in)famous Fermat Conjecture. This states that the equation $x^n + y^n = z^n$ has no solution in integers x, y, and z for any integer n greater than 2. We will take up a longer look at the Wolfskehl Prize in a later chapter of this volume, and its final awarding to Andrew Wiles of Princeton University in 1997.

Without a doubt the most influential list of problems ever put together was the collection of 23 problems announced by David Hilbert in his address to the International Mathematical Congress in Paris in 1900. In this list, whose solutions Hilbert felt were important for the future development of mathematics, we find both specific problems and "problems," these latter being more like problem areas or research programs than specific questions. An example of the former is the famed Tenth Problem, which asks for a single method for determining the solvability or nonsolvability of any given Diophantine equation (basically a polynomial equation with integer coefficients for which we demand integer solutions). We present the solution to this problem showing that no such method can possibly exist in a later chapter of this volume. In the latter category of "problems" we find the last problem on Hilbert's list, Problem 23, which seeks "Further development of the methods of the calculus of variations." But, of course, one might well ask the same for every branch of mathematics! So in this sense Hilbert's list is very idiosyncratic, and certainly not all problems on the list are of equal import.

Interestingly, missing from Hilbert's list are two of the most spectacular problems of the time: the Fermat Conjecture and the N-Body Problem, both noted earlier. Even more strange is that both in his actual lecture and in the full version of the published paper, Hilbert mentioned both as problems in his preamble but omitted them from the list of problems itself. So are there really 25 problems in all?

It's even more amusing to read accounts of the lecture by attendees. For example, Charlotte Angas Scott wrote in the *Bulletin of the American Mathematical Society* that the reaction after Hilbert's lecture was "a rather desultory discussion." In fact, only two comments were made. The first, by the Italian Giuseppe Peano, who remarked that the Second Problem on the consistency of arithmetic was already essentially solved by his colleagues. The second comment, from the German mathematician Rudolf Mehmke, made a point about numerical methods that bore on the Thirteenth Problem on resolving polynomial equations of the fifth degree.

Maybe it was Hilbert's delivery that was partly to blame for this feeble reaction. Scott opined that "the presentation of papers is usually shockingly bad," with monotonic utterances exuding boredom; she gave no names, but hinted that eminent ones were not excluded. Shades of modern congresses and symposia! Even Hilbert himself did not mention his lecture two weeks after the Congress when he reported on the meeting in a letter to his colleague, Adolf Hurwicz, who had not attended.

Hilbert seems to have conceived his lecture as a counter to the rather bland advocacy of the importance of applied mathematics made by Poincaré at the First International Congress of Mathematicians in Zürich in 1897. But the brilliance with which he focused on several specific problems (for example, the Continuum Hypothesis (Problem #1) and the solvability of Diophantine equations (Problem #10)) seems to have brought some snow-blindness to the estimation by later mathematicians of the value of the entire collection. But to his defense, Hilbert admitted that he had made a personal selection of problems. He said "they are only examples of problems," though he also claimed that they showed "how extensive is the mathematical science of today." So it's interesting to ponder whether the glamor given to his selection may have distorted priorities in the development of twentieth-century mathematics. Be that as it may, the one thing that cannot be denied is the prestige and status bestowed on mathematicians who manage to crack one of these "Big 23." In fact, no less than three of the five mountaintops considered here are from Hilbert's list (and that could be extended to four if we consider the Fermat Conjecture to be admitted to Hilbert's pantheon).

If good problems are the lifeblood of mathematics, then so are good answers. What do we mean by an answer? At first hearing this seems obvious. We mean a full-fledged, airtight deductive proof that a certain assertion follows logically from the axioms of the logical framework within which we are working. Well, things are never as simple as they seem—even when it comes to mathematical proof. Let's have a brief look at why.

PROOF IS A MANY SPLENDORED THING

Paul Erdös, the brilliant Hungarian mathematician who collaborated with more people than just about anyone who ever lived, reckoned that up in heaven God had a book that contained all the best mathematical proofs. If Erdös was really impressed by a proof, he declared it to be "from the Book." In his view, the mathematician's job is to sneak a look over God's shoulder and pass on the beauty of His creations to the rest of the world.

As will be seen in the stories in this volume, it seems as if this simple, elegant approach is just one of a number of possible narrative lines for mathematical proofs. For example, the formal proof by Andrew Wiles of Fermat's Conjecture, recounted in a later chapter, runs to something over 200 printed pages. Similarly, the Classification Theorem for Finite Groups has a proof that extends over dozens of journal articles that must total at least a thousand pages or more. One might wonder whether these mammoth proofs are necessary. And are they so vast only because mathematicians are too stupid to find really short, clever versions written in The Book?

One answer is to that there is nothing to say that every short, simple and true statement should have a short, simple, and true proof. In fact, the Austrian-American mathematician Kurt Gödel showed that short statements can sometimes have very long proofs. It's just that there is no way to know ahead of time which statements these are!

Besides the two narrative styles of proof—short and simple and long and complex—there is a third style, the computer-assisted proof. Two of the chapters of this volume, the accounts of the Four-Color Conjecture and the Kepler Conjecture, are good examples of this type of proof. Take the Kepler Conjecture. In 1611, Johannes Kepler considered the way that spheres could be packed together. He concluded that the way that packed as many balls as possible into a given region was the one that grocers use to stack grapefruits and oranges. This packing involves making a flat layer in a honeycomb pattern, then stack another layer on top, sitting in the depressions of the first layer, and continue in this manner forever.

Even the two-dimensional version of this problem, which shows that a honeycomb pattern is the most efficient way to pack equal circles in a plane, wasn't proved until 1947 by Hungarian mathematician, Lazlo Fejes Tóth. In 1998, though, Thomas Hales of the University of Michigan announced a computer-assisted proof that involved hundreds of pages of mathematics plus a vast amount of supporting computer calculations. His approach was to write down a list of all the possible ways to arrange suitable small clusters of spheres, then prove that whenever the cluster is not what you find in the Kepler pattern it can be compressed by rearranging the spheres slightly. This led to the conclusion that the only incompressible arrangement—the one that fills space most efficiently—is the one that Kepler claimed was best. In fact, this was the same approach that Fejes Tóth used to solve the planar version of the problem, where he needed to list about 50 possibilities. Hales had to deal with thousands, and the computer had to verify an enormous list of inequalities that occupied over 3 gigabytes of computer memory.

Computer-assisted proofs raise a number of questions regarding taste, creativity, technique, and philosophy. We will consider these in more detail in later chapters. But it's worth mentioning at least a few of the issues briefly here. Some philosophers of mathematics feel that the brute-force methods of computer proofs are not really proofs at all, at least not in the traditional sense. Others say that this kind of massive but routine exercise is what computers do best, and humans do poorly. So if a computer and a human both carry out a huge calculation and come to different answers, the smart money goes on the computer.

But these type of proofs are not without their charm and elegance. You have to be clever about how you set up the problem for the computer to attack it. What's more, once the computer has helped establish that the assertion you're trying to prove is indeed correct, you can try to find a more elegant way to solve it. It's a known fact that it's a lot easier to prove something if you already know it's right. On the other hand, this certainly does not mean that eventually mathematicians will find God's proofs for the Fermat or the Four-Color Conjectures. Perhaps there are no proofs of these theorems in The Book. Remember Gödel's discovery that some proofs simply have to be long. Perhaps the Four-Color Theorem is an example of this. Certainly if you want to use the current approach, which involves finding a list of unavoidable configurations and then eliminating them one at a time by some "shrinking" process, then nothing radically shorter is possible. The increasingly widespread use of computers in mathematics raises another set of interesting questions about mathematical proof, as well.

The English word "prove"—as its Old French and Latin ancestors—has two basic meanings: to try or test, and to establish beyond doubt. The first meaning is by now mostly archaic, although it survives in old adages, such as "the exception proves the rule" and "the proof is in the pudding." Strangely, in mathematics, where the second meaning is the order of the day, the first meaning also survives. Most mathematicians spend a lot of their time thinking about and analyzing particular examples. Gauss's notebooks attest to his statement that his way of arriving at mathematical truths was "through systematic experimentation." And it's probably true that most significant advances in mathematics come from experimenting with examples. A good modern example is the discovery of the tree structure of Julia sets by Duoady and Hubbard. This was first observed by looking at pictures produced by computers and was then proved by formal arguments.

It's of considerable concern that such empirical considerations are almost always completely excluded from the published record. What we see in print is a snowstorm of logical argumentation that only the most dedicated will attempt to follow. One might ask if this is to the advantage of the mathematical community? Should mathematics be presented as some sort of magic conjured out of thin air by extraordinary people when it is actually the result of hard work and intuition built on the study of many special cases? In essence, mathematics is a experimental activity and the published record of mathematical results should reflect this fact.

The example from Gauss's notebooks shows that experimental mathematics has always been around. So the introduction of the computer into mathematics is more of a quantitative change than a qualitative one. But in another way it has been quite revolutionary, affecting at a deep level the philosophy of the professional mathematician.

Computers in mathematics reinforce the constructivist view that mathematical objects exist solely in virtue of a procedure for actually fabricating them. This point of view, in turn, increases the use of the computer, since the computer program that produces a tree structure for a Julia set, for instance, is that very procedure.

The role of the computer in suggesting conjectures and enriching our understanding of abstract concepts by means of examples and visualization is a healthy and welcome development in mathematics. Far from undermining rigor, the use of computers in mathematics will enhance the field in many ways. Mathematicians who write complicated computer programs soon realize that subjecting lines of code to the usual techniques of mathematical analysis and proof often reveals faults in the programs. Thus, programming can enhance an appreciation of why mathematicians regard proof as important. Moreover, the use of computers gives mathematicians another view of mathematical reality and another tool for investigating the correctness of a piece of mathematics through examining examples. Finally, use of the computer will strengthen the trend toward constructivism, helping to recast mathematics on a more solid foundation.

CHAPTER ONE

HILBERT'S TENTH PROBLEM

25 OF HILBERT'S BEST

David Hilbert was one of the super giants of twentieth-century mathematics. His work in logic, geometry, algebra, functional analysis, number theory, and a dozen other areas left a mark on virtually every area of mathematics known in his time. As recognition of his stature in the community, Hilbert was invited to deliver one of the key addresses at the International Congress of Mathematicians in Paris in 1900. Usually such talks give a broad retrospective on some topic, tracing developments in the past that led to where the field stands at present. But Hilbert had another strategy in mind. As the entire meeting had been shifted by one year in order to take place at the turn of the century, Hilbert decided to use this occasion to look *forward* rather than backward, and to give a list of problems whose solutions he felt would be pivotal for the development of mathematics in the coming century. In the lecture itself, Hilbert presented ten such problems, starting with the famed Continuum Problem discussed elsewhere in this book. He later augmented this collection of problems, adding 13 problems to the published version of his lecture and two more in his introductory remarks to the paper. This list of 25 problems has fulfilled Hilbert's expectations in every possible way, serving as highly visible mountaintops, the scaling of which ensures fame and recognition for any mathematician lucky enough and skilled enough to reach them. Certainly one of the most challenging problems on this list is the tenth one, which is the only one to have subsequently been enshrined in the lexicon of mathematics solely by its position on the list. Let's see what Hilbert's Tenth Problem is all about.

As stated by Hilbert (in English translation) in the written version of his lecture, the Tenth Problem is

> **Determination of the Solvability of a Diophantine Equation.** Given a Diophantine equation with any number of unknown quantities and with rational integral numerical coefficients: *To devise a process according to which it can be determined by a finite number of operations whether the equation is solvable in rational integers.*

So what does all this mean? Well, first of all it deals with the solution of a particular type of equation called a *Diophantine* equation. This is a polynomial equation in many variables, which means that the unknown quantities are raised to integer powers. Examples of such equations are the Pythagorean equation $x^2+y^2 = z^2$ and the linear equation $ax+by = c$ for unknowns x, y, and z. Here the quantities a and b are the coefficients in the second equation, while the coefficients are all the number 1 in the first one. Hilbert was interested in these types of equations where the coefficients were restricted to be rational numbers, the ratio of integers, not real numbers like π or $\sqrt{2}$. So the coefficients in the Tenth Problem can only be quantities like 2/3, 1, or 75/79. Now what does the term "Diophantine" mean?

Diophantus of Alexandria lived around two millennia ago. He was famous for a textbook he wrote on algebra, in which he considered polynomial equations with integer coefficients, in which it was required that the solutions, the values of $x, y, z, \ldots$ are also integers. Such equations with this condition on the solutions have subsequently come to be termed Diophantine equations. Thus, the terminology refers more to the kind of solution that one seeks than to the equation itself. So an equation like the Pythagorean equation above is simply an "equation" if real-number solutions are sought. But it is a Diophantine equation if we are looking for integer solutions such as $x = 3, y = 4, z = 5$.

Solving Diophantine equations is rather different from solving the same equation in real numbers. For example, the quadratic equation

$$x^2 + y^2 = 2$$

has the infinitude of real solutions $x = r, y = s$, where

$$s = +\sqrt{2 - r^2},$$

and r is any real number between $-\sqrt{2}$ and $+\sqrt{2}$. On the other hand, considered as a Diophantine equation there are only the four solutions $x = \pm 1, y = \pm 1$. Furthermore, changing the equation slightly to $x^2 + y^2 = 3$, there are still an infinite number of real solutions—but no integer solutions at all. So as a Diophantine equation this latter version has no solution; it's unsolvable. The Tenth Problem focuses on this question of solvability of such equations. In particular, Hilbert asks for a single "process" that can be used to decide whether any polynomial equation with rational coefficients has a solution in integers. That's the problem.

Hilbert's statement of the problem called for a solution in integers, that is, positive or negative numbers. One can also state the matter in terms of natural numbers (including 0). These are two very different problems. For instance, the equation

$$(x+1)^3 + (y+1)^3 = (z+1)^3,$$

has infinitely many integer solutions of the form $x = -1, y = z$. But it has no solutions in natural numbers.

On the other hand, let

$$C(x_1, x_2, \ldots, x_n) = 0,$$

be an arbitrary Diophantine equation. Suppose we are seeking its solution in integers. Consider the related equation

$$D(p_1 - q_1, p_2 - q_2, \ldots, p_n - q_n) = 0.$$

It's clear that any solution of this equation in natural numbers $p_1, p_2, \ldots, p_n, q_1, q_2, \ldots, q_n$ yields the solution of the first equation

$$x_1 = p_1 - q_1, x_2 = p_2 - q_2, \ldots, x_n = p_n - q_n$$

in integers $x_1, x_2, \ldots, x_n$. Moreover, for any $x_1, x_2, \ldots, x_n$ forming a solution of the first equation in integers, we can find natural numbers $p_1, p_2, \ldots, p_n, q_1, q_2, \ldots, q_n$ satisfying the second equation.

What this argument shows is that the problem of solving the first equation in integers reduces to the problem of the solvability of the second equation in natural numbers. So there is no loss of generality in considering the problem of the solvability of Diophantine equations in integers as the question of solvability in natural numbers. In short, the two problems are very different *for a particular equation* but they are equivalent when considering the entire *class* of Diophantine equations. For a variety of technical reasons, it's a bit easier to work with variables that range over the natural numbers instead of the integers. So we will assume that that is the case in what follows unless stated explicitly to the contrary.

Notice that the Tenth Problem asked for a *single* method that could be applied to *any* polynomial equation. Since the time of Diophantus, mathematicians had found solutions for lots of Diophantine equations and proved others unsolvable. But for different types of equations different methods had to be used to obtain these results. In the Tenth Problem, Hilbert asked for a *universal* method for determining solvability of Diophantine equations.

In today's terminology, the Tenth Problem is what's called a *decision problem.* It is a problem consisting of an infinite number of questions (the infinite number of possible equations), each of which requires a Yes or No answer. The core of a decision problem is always to find a single method that can be applied to each question, and that will always give the correct answer for each individual problem. It's of passing interest to remark that the Tenth Problem is the only decision problem on Hilbert's list of 25 "big ones" for the coming century. One of the reasons this problem occupies such an important position in twentieth-century mathematics is that it served to focus attention on the idea of an *algorithm,* certainly one of the key concepts underlying much of modern computing and the overall philosophy of mathematics.

THE SOLVABLE AND THE IMPOSSIBLE

In today's language, what Hilbert had in mind when he spoke of a "process" to determine the solvability of a Diophantine equation was an algorithm. Here's a simple example to fix the idea.

Consider the linear Diophantine equation

$$ax - b = 0$$

where a and b are integers. It's clear that this equation has an integer solution if and only if a divides b exactly, in which case the solution is the integer b/a. So it's no problem to see what the program would be to test such an equation for integer solutions. It would consist of the two steps:

1. Compute the quantity $q = b/a$.

2. If q is an integer, then the equation has the integer solution q; otherwise, the equation is not solvable.

Now what about a linear equation in two variables, x and y:

$$ax + by = c,$$

where a, b, and c are integers? Here again there is a simple procedure for determining whether there are integers x and y solving this equation. The program in this case is:

1. Compute d, the highest common factor of a and b.
2. If d divides c exactly, then there is an integer solution; otherwise, there is no integer solution.

At this juncture it's important to mention that having a method to determine whether an equation has an integer solution is *not* the same thing as having a method for finding that solution if it exists. But it turns out that for linear Diophantine equations, there is such a procedure for actually finding the solution. It's called the *Euclidean algorithm,* which involves finding the highest common factor of two numbers, as called for in step 1 above. This step-by-step method appeared in Book VII of Euclid's *Elements,* which is why it's called the Euclidean algorithm today. But what *exactly* is an algorithm? That question leads us into one of the most exciting developments of twentieth-century mathematics, the theory of computation.

What is a computation? Oddly enough, despite the fact that humans have been calculating (and there have been calculating humans!) for thousands of years, a proper scientific answer to this seemingly straightforward query was not forthcoming until 1935. In that year, Alan Turing was a student at Cambridge University sitting in on a course of lectures in mathematical logic given by the esteemed logician Max Newman. A central theme of the course was the issue of whether or not there could exist a finite set of rules, in effect a *mechanism,* that would settle the truth or falsity of every possible statement about numbers that could be made in, say, the language of Russell and Whitehead's (in)famous work, *Principia Mathematica.* In short, the question was whether there was a machine into which we could feed any statement about numbers so that after a finite amount of time the machine would spit out the verdict on the statement, TRUE or FALSE.

Turing's speculations about what it would mean to have such a mechanical procedure, or *effective process*, for solving this famous Decision Problem led him to develop a mathematical type of computer. This abstract gadget, the Turing machine, provided the first completely satisfactory answer to what it means to carry out a computation.

Turing took the commonsense view of looking at what a human being actually does when carrying out a computation. As it turns out, the distilled essence of computing comes down to the rote following of a set of rules. So, for example, if you want to calculate the square root of 2, you might employ the following rule for creating a set of numbers $\{x_i\}$ that will (hopefully) converge to the quantity $\sqrt{2}$: $x_{n+1} = (x_n/2) + (1/x_n)$. Starting with the initial approximation (i.e., guess) $x_0 = 1$, this rule generates the successively better approximations $x_1 = 3/2 = 1.5$, $x_2 = 17/12 = 1.4166$, $x_3 = 577/408 = 1.4142$. So after just three steps, we have the desired answer correct to four significant figures. For our purposes here, what's important about this so-called Newton-Raphson method for calculating the square root of 2 is not the rapid rate of convergence, but that the procedure represents a purely mechanical, step-by-step process (technically, an *algorithm*) for finding the desired quantity.

The fact that every step in such a procedure is completely and explicitly specified led Turing to believe that it would be possible to construct a machine to carry out the computations. Once the algorithm and the starting point are given to the machine, computation of the sequence of results becomes a purely mechanical matter, involving no judgment calls or interventions by humans along the way. But it would require a special type of machine to accomplish this computational task; not just any mechanical device will do. A large part of Turing's genius was to show that the very primitive type of abstract computing machine he invented is actually the most general type of computer imaginable. In fact, every real-life computer that's ever been built is just a special case that materially embodies the machine that Turing dreamed up. This result is so central to understanding the limitations of machines that it's worth our while to take a few pages to describe it in more detail.

MAGIC MACHINES

A Turing machine consists of two components: (1) an infinitely long tape ruled off into squares that can each contain one of a finite set of symbols, and (2) a scanning head that can be in one of a finite number of states or configurations at each step of the computational process. The head

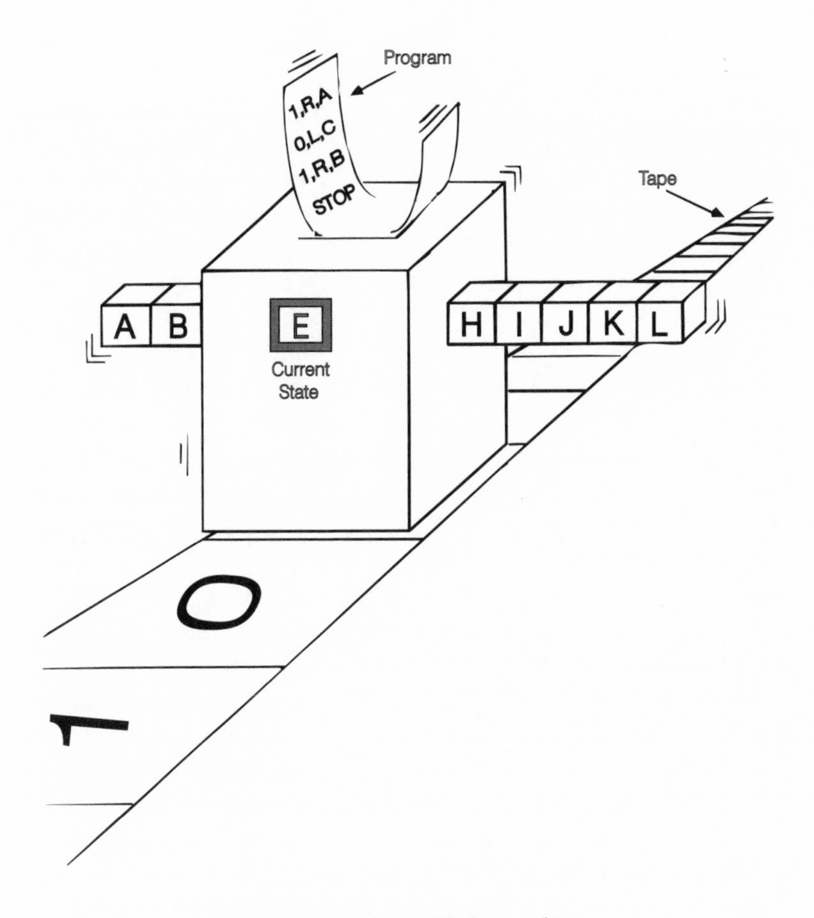

Figure 1.1. A 12-state Turing machine.

can read the squares on the tape and write one of the symbols onto each square. The behavior of the Turing machine is controlled by an algorithm, or what we now call a *program.* The program is composed of a finite number of instructions, each of which is selected from the following set of possibilities: change or retain the current state of the scanning head; print a new symbol or keep the old symbol on the current square; move left or right one square; stop. That's it. Just seven simple possibilities. The overall situation is depicted in Figure 1.1 for a Turing machine having 12 internal states labeled A through L. Which of the seven possible actions the head takes at any step of the process is determined by the current state of the head and what it reads on the square it's currently scanning. But rather than continuing to speak in these abstract terms, it's simpler to run through an example in order to get the hang of how such a device operates.

Assume we have a Turing machine with three internal states, *A*, *B*, and *C*, and that the symbols that can be written on the tape are just the two integers 0 and 1. Now suppose we want to use this machine to carry out the addition of two whole numbers. For definiteness, we'll represent the integer *n* by a string of *n* consecutive 1s on the tape. The program shown in Table 1.1 serves to add any two whole numbers using this 3-state Turing machine.

Table 1.1. A Turing machine program for addition.

	Symbol Read	
State	1	0
A	1, R, A	1, R, B
B	1, R, B	0, L, C
C	0, STOP	STOP

The reader should interpret the table entries in the following way: the first entry is the symbol the head should print, the second element is the direction the head should move, R(ight) or L(eft), while the final element is the state the head should then move into. Note that the machine stops as soon as the head goes into state C. Let's see how it works for the specific case of adding the numbers 2 and 5.

Since our interest is in using the machine to add 2 and 5, we place two 1s and five 1s on an otherwise blank (all 0s) input tape, separating them by a 0 to indicate that they are two distinct numbers. Thus the machine begins by reading the input tape

...	0	0	0	1	1	0	1	1	1	1	1	0	0	0	0	...

By convention, we assume the head starts in state A and reads the first nonzero symbol on the left. Since this symbol is a 1, the program tells the machine to print a 1 on the square and move to the right, retaining its internal state A. The head is still in state A and the current symbol read is again a 1, so the machine repeats the previous step and moves one square farther to the right. Now, for a change, the head reads a 0. The program tells the machine to print a 1, move to the right, and switch to state B. I'll leave it to the reader to complete the remaining steps of the program, verifying that when the machine finally halts, the tape ends up looking just like the input tape above, except with the 0 separating 2 and 5 having been eliminated (i.e., the tape will have seven 1s in a row, as required).

Before looking at the revolutionary implications of Turing's idea, let me pause here to emphasize that Turing machines are definitely not machines in the everyday sense of being material devices. Rather they are "paper computers," completely specified by their programs. Thus, when we use the term *machine* in what follows, the reader should read *program* or *algorithm* (i.e., software) and put all notions of hardware out of sight and out of mind. This abuse of the term *machine* should have been clear from Turing's idea of an *infinite* storage tape, but it's important to make the distinction as clear as possible: Turing machine = program. Period.

Modern computing devices, even home computers like the one I'm using to write this book, look vastly more complicated and powerful in their computational power than a Turing machine with its handful of internal states and very circumscribed repertoire of scanning-head actions. Nevertheless, this turns out not to be the case, and a large measure of credit is due to Turing for recognizing that *any* algorithm (i.e., program) executable on *any* computing machine—idealized or otherwise—can also be carried out on a particular version of his machine, termed a *universal Turing machine* (or UTM for short). So except for the speed of the computation, which definitely *is* hardware-dependent, there's no computation that my machine (or anyone else's) can do that can't be done with a UTM.

To specify his UTM, Turing realized that not only the input data of the problem, but also the program itself could be coded by a series of 0s and 1s. Consequently, we can regard the program as another kind of input data, writing it onto the tape along with the data it is to process. Table 1.2 shows one of the many ways this coding can be done.

With this key insight at hand, Turing constructed a program that could simulate the action of any other program $\mathcal{P}$ when given $\mathcal{P}$ as part of its input (i.e., he created a UTM). The operation of a UTM is simplicity itself.

Suppose we have a particular Turing machine specified by the program $\mathcal{P}$. Since a Turing machine is completely determined by its program, all we need do is feed the program $\mathcal{P}$ into the UTM along with the input data. Thereafter the UTM will simulate the action of $\mathcal{P}$ on the data; there will be no recognizable difference between running the program $\mathcal{P}$ on the original machine or having the UTM pretend it *is* the Turing machine $\mathcal{P}$.

Table 1.2. A coding scheme for the Turing machine language.

Program statement	**Code**
PRINT 0	000
PRINT 1	001
GO RIGHT	010
GO LEFT	011
GO TO STEP i IF THE CURRENT SQUARE CONTAINS 0	$101\underbrace{00\ldots\ldots}_{i \text{ repetitions}}01$
GO TO STEP i IF THE CURRENT SQUARE CONTAINS 1	$110\underbrace{11\ldots\ldots}_{i \text{ repetitions}}10$
STOP	100

What's important about the Turing machine from a theoretical point of view is that it represents a formal mathematical object. So with the invention of the Turing machine, for the first time we had a well-defined notion of what it means to compute something. But this then raises the question: What exactly can we compute? In particular, is there a suitable Turing machine that will compute every number? Or do there exist numbers that are forever beyond the bounds of computation? Turing himself addressed this problem of computability in his trail-blazing 1936 paper, in which he introduced the Turing machine as a way of answering these fundamental questions.

First of all, let's be clear on what we mean by a number being computable. Put simply, an integer n is said to be *computable* if there is a Turing machine that, starting with a tape containing all 0s, will stop after a finite number of steps with the tape then containing a string of n 1s and all the rest 0s. The case of computing a real number is a bit trickier since most real numbers consist of an infinite number of digits. So we call a real number computable if there is a Turing machine that will successively print out the digits of the number, one after the other. Of course, in this case the machine will generally run on forever. With these definitions in hand, let's look at the limitations on our ability to compute numbers.

It's an easy exercise to show that for a two-symbol Turing machine with n possible states of the reading head, there are exactly $(4n+4)^{2n}$ distinct programs that can be written. This means that an n-state machine can compute at most this many numbers. Letting n take on the values $n = 1, 2, 3, \ldots$, we conclude that Turing machines can calculate at most a *countable* set of numbers, that is, a set whose elements can be put into a one-to-one correspondence with a subset of the positive integers (the "counting" numbers). But there are uncountably many real numbers; hence, we come to the perhaps surprising result that the vast majority of real numbers are not computable.

This counting argument is one way to show the existence of uncomputable numbers, albeit a somewhat indirect one. Turing himself used a more direct procedure based upon what's known as *Cantor's diagonal argument.* It goes like this. Consider the following listing of names Smith, Otway, Arquette, Bethel, Bellman, and Imhoff. Now take the first letter of the first name and advance it alphabetically by one position. This gives a T. Then do the same for the second letter of the second name, the third letter of the third name and so on. The result is "Turing." It's clear, I think, that the name Turing could not have been on the original list, since it must differ from each entry on that list by at least one letter.

Turing's argument for the existence of uncomputable numbers follows the same line of reasoning. Suppose you list all computable numbers, written out by their decimal expansions even though such a list will be infinitely long. Now advance the first digit of the first number, the second digit of the second number and, in general, the kth digit of the kth number. In this way we create a new number. This number cannot have been on the original list, since it differs in at least one position from every number on that list. But by definition, the list contains all computable numbers. Hence the new number must be uncomputable.

From the foregoing arguments, we see that uncomputable numbers are not *rara avis* in the arithmetic aviary. Quite the contrary, in fact, as it's the computable numbers that are the exception rather than the rule. This surprising fact shows that all the numbers we deal with in our everyday personal and professional lives, which by their very nature must be computable, form but a microscopically small subset of the set of all possible numbers. The overwhelming majority of numbers lie in a realm that's impossible to reach by following the rules of any type of computing machine. Now to firmly fix the difference between something's existing and our being able to compute that same something, let's consider a game, the Turing Machine Game.

The Turing Machine Game

Assume there are two players, called rather unimaginatively A and B. These players take turns choosing positive integers as follows:

Step 1: Player A chooses a number n.
Step 2: Knowing n, Player B picks a number m.
Step 3: Knowing m, Player A selects a number k.

Player A wins if there is some n-state Turing machine that halts in exactly $m+k$ steps when started on a tape containing all 0s. Otherwise, Player B wins. It's fairly clear that this is a game of finite duration, since once the players have chosen their integers, all we need do is list the $(4n + 4)^{2n}$ Turing machines having n states, and run each of them for exactly $m+k$ steps to determine a winner.

It's a well-known fact from game theory that any game of fixed, finite duration is determined, in the sense that there is a winning strategy for one of the players. In this case, it's Player B. Nevertheless, the Turing Machine Game is nontrivial to play, since neither player has an *algorithm* (i.e., a computable strategy) for winning the game. The proof of this fact relies on showing that *any* winning strategy involves computing a function whose values grow faster than those of a particular uncomputable function. Hence, this new function must also be uncomputable. The reader is referred to the papers cited in the chapter references for further details of this proof.

From our definition of computability, it's clear that you haven't really computed anything until the computational process terminates—even in the case of real numbers, where any finite computation generally yields only an approximation to the number you're trying to compute. This simple observation leads to a key question in the theory of computation: is there a general procedure (i.e., an algorithm) that will tell us *in advance* whether or not a particular program will halt after a finite number of steps? In other words, given any Turing machine program $\mathcal{P}$ and a set of input data $\mathcal{I}$, is there a single program that accepts $\mathcal{P}$ and $\mathcal{I}$ and that will tell us whether or not $\mathcal{P}$ will halt after a finite number of steps when processing the data $\mathcal{I}$? Note carefully that what we're asking for here is a *single* program that will work in *all* cases. This is the famous *Halting Problem.*

To see that the question is far from trivial, suppose we have a program $\mathcal{P}$ that reads a Turing machine tape and stops when it comes to the first 1. So in essence the program says, "Keep reading until you come to a 1, then stop." In this case, the input data $\mathcal{I}$ consisting entirely of 1s would result in the program stopping after the first step. On the other hand, if the input data were all 0s, then the program would never stop. Of course, in this situation we have a clear-cut procedure for deciding whether or not the program will halt when processing some input tape: the program will stop if and only if the input tape contains even a single 1; otherwise, the program will run on forever. So here's an example of a halting rule that works for any data set processed by this especially primitive program.

Unfortunately, most real computer programs are vastly more complicated than this, and it's far from clear by simple inspection of the program what kinds of quantities will be computed as the program goes about its business. After all, if we knew what the program was going to compute at each step, we wouldn't have to run the program. Moreover, the stopping rule for real programs is almost always an implicit rule of the foregoing sort, saying something like "If such and such a quantity satisfying this or that condition appears, stop; otherwise, keep computing." The essence of the Halting Problem is to ask if there exists any *effective procedure* that can be applied to the program and its input data to tell beforehand whether or not the program's stopping condition will ever be satisfied. In 1936, Turing settled the matter once and for all in the negative: given a program $\mathcal{P}$ and an input data set $\mathcal{I}$, there is no way in general to say if $\mathcal{P}$ will ever finish processing the input $\mathcal{I}$.

The notion of a Turing machine finally put the idea of a computation on a solid mathematical footing, enabling us to pass from the vague, intuitive idea of an effective process to the precise, mathematically well-defined notion of an algorithm. In fact, Turing's work, along with that of the American logician, Alonzo Church, forms the basis for what has come to be called the

Turing-Church Thesis. Every effective process is implementable by running a suitable program on a UTM.

The key message of the Turing-Church Thesis is the assertion that any computable quantity can be computed by a suitable Turing machine. This claim is called a thesis and not a theorem because it's not really susceptible to proof. Rather, it's more in the nature of a definition, or a proposal, suggesting that we agree to equate our informal idea of carrying out a computation with the formal mathematical idea of a Turing machine.

To bring this point home more forcefully, it's helpful to draw an analogy between a Turing machine and a typewriter. A typewriter is also a primitive device, allowing us to print sequences of symbols on a piece of paper that is potentially infinite in extent. A typewriter also has only a finite number of states that it can be in: upper and lower case letters, red or black ribbon, different symbol balls and so on. Yet despite these limitations, any typewriter can be used to type *The Canterbury Tales, Alice in Wonderland,* or any other string of symbols. Of course, it might take a Chaucer or a Lewis Carroll to tell the machine what to do. But it can be done. By way of analogy, it might take a very skilled programmer to tell the Turing machine how to solve difficult computational problems. But, says the Turing-Church Thesis, the basic model—the Turing machine—suffices for every type of problem that is at all solvable by carrying out a computation.

While it's not central to our story, it's of more than passing interest to note that Turing's solution of the Halting Problem is completely equivalent to Gödel's 1931 result on the limitations of formal logical systems. Here are statements of the two results:

Gödel's Theorem. For any consistent formal system $\mathcal{F}$ purporting to settle, that is, prove or disprove, all statements of arithmetic, there exists an arithmetical proposition that can be neither proved nor disproved in this system. Therefore, the formal system $\mathcal{F}$ is incomplete.

The Halting Theorem. For any Turing machine program $\mathcal{H}$ purporting to settle the halting or nonhalting of all Turing machine programs, there exists a program $\mathcal{P}$ and input data $\mathcal{I}$ such that the program $\mathcal{H}$ cannot determine whether or not $\mathcal{P}$ will halt when processing the data $\mathcal{I}$.

When placed side by side in this fashion it becomes fairly evident, I think, that the Halting Theorem is simply Gödel's Theorem expressed in terms of computing machines and programs instead of in the language of logical deductive systems. Now, let's get back to the Tenth Problem. For that we need a couple of new ideas, the notion of a *computable* set and that of a *listable* set of integers.

Following Turing's work in setting the idea of what we mean by a "computation" on a solid mathematical footing, it was finally possible to precisely define what is and is not computable. In particular, it turns out that a crucial notion of the Tenth Problem is that of a *computable* set of integers. A set S of integers is called computable if there is a Turing machine program for determining which numbers are in S and which are not. The program would be given an integer as input, and stop after a finite number of steps with an output of 1 if the number is in S and halt with an output of 0 if the number is not in S.

Notice that this definition of computable set requires that the Turing machine program always halt with a definite answer regardless of what integer is given at the outset. But we're all familiar with programs that run on forever, going into an infinite logical loop or a search for data that doesn't exist. A weaker notion than computable that allows for such never-ending calculations is that of a *listable* set of integers (technically, a *recursively enumerable* set). A set of integers S is called listable if there is a Turing machine program that, given any integer as input, halts with output 1 if and only if the integer belongs to the set S. If the integer does not belong to S, the program may or may not halt. So if you run the program on a given integer K, if K happens to be a member of S then the program will eventually give you this information. But if K is not a member of S, you may never know it—the computation may go on forever.

A little thought shows that there is a close relationship between a set being computable and its being listable. A set S of integers is computable if and only if both S and its complement, $\bar{S}$, are listable (Note: the complement of S is all integers that are *not* in S). But the two notions are definitely not the same. Every computable set is listable, but there are listable sets that are not computable. For a proof of this fact, the reader is referred to the reference section for this chapter.

Besides individual Diophantine equations, we can also consider *families* of such equations. Such a family takes the form

$$D(a_1, a_2, \ldots, a_n, x_1, x_2, \ldots, x_m) = 0, \qquad (*)$$

Where D is a polynomial with integer coefficients. Here the variables in the equation are split into two groups:

- *parameters* $a_1, a_2, \ldots, a_n$;
- *unknowns* $x_1, x_2, \ldots, x_m$.

Here we assume the parameters also can only take on values in the set of natural numbers.

For some choices of the values of the parameters $a_1, a_2, \ldots, a_n$ the equation has a solution in integers for the unknowns $x_1, x_2, \ldots, x_m$, while for other choices of the parameters there is no such solution. Thus, we can consider the set $\mathcal{M}$ of all n-tuples of parameters $(a_1, a_2, \ldots, a_n)$ for which the equation (∗) has a solution. Such sets are called *Diophantine sets,* since their elements are defined by the solvability of the Diophantine equation (∗).

Here is an example of a Diophantine set:

- *the set of all squares,* which are given by the solution of the Diophantine equation

$$a - x^2 = 0.$$

Here all values of the single parameter a that are squares are in the set $\mathcal{M}$, while all other positive natural numbers are not in $\mathcal{M}$.

It's rather less evident—but still true—that the set of all numbers that are not squares is also a Diophantine set associated with the equation

$$(a - z^2 - x - 1)^2 + \left((z + 1)^2 - a - y - 1\right)^2 = 0.$$

So all values of the parameter a for which this equation has positive integer solutions x, y, z are *not* squares.

It's rather natural to ask about how to characterize the entire class of Diophantine sets. In other words, is there some set of conditions that is both necessary and sufficient for a particular set of integers to be Diophantine? There is one clear necessary condition, that is easily seen when we look at Diophantine sets from a computational point of view. Suppose we are given the family of Diophantine equations (∗). It is possible to effectively list all n-tuples from the Diophantine set $\mathcal{M}$ represented by this equation. Here's how. In some systematic fashion, look over all $(n+m)$-tuples of possible values of the variables $a_1, a_2, \ldots, a_n, x_1, x_2, \ldots, x_m$, and check to see when equality holds in equation (∗). When it does, put that n-tuple $(a_1, a_2, \ldots, a_n)$ into the set of elements of $\mathcal{M}$. In this way, every n-tuple from $\mathcal{M}$ will appear on the list sooner or later (and perhaps many times). What this procedure shows is that for a set $\mathcal{M}$ to be Diophantine, it must be listable. The modern era of attack on Hilbert's Tenth Problem began when logician Martin Davis conjectured that this condition is also *sufficient.* In short, Davis said that a set is Diophantine if and only if it is listable. Let's now turn to a detailed examination as to why this conjecture was important, and how it leads to the resolution of the Tenth Problem.

ASSAULT ON THE TENTH

While an undergraduate student in mathematics at the City College of New York shortly after World War II, Martin Davis read in one of his teacher Emil Post's articles that Hilbert's Tenth Problem "begs for an unsolvability proof." Post was one of the pioneers of logic, and in the early 1920s developed a scheme to formalize the notion of a computation very similar to what Turing proposed some years later. So Davis was sitting at the feet of a master, someone who knew whereof he spoke. With the confidence inspired by Post's assertion, Davis set out to prove that the magic algorithm sought by Hilbert was a chimera; it simply did not exist. His strategy consisted of the following chain of steps:

1. Show that for every listable set S, there is a polynomial $P_S(x, y_1, y_2, \ldots, y_n)$ with integer coefficients, such that a positive number k belongs to the set S if and only if the equation

$$P_S(k, y_1, y_2, \ldots, y_n) = 0$$

has a solution in positive integers, that is, it is Diophantine. The degree of P_S turns out to be unimportant, as does the number of variables n. What *is* important is that such a polynomial can be associated with every listable set.

2. Now, argued Davis, suppose there did exist an algorithm of the type sought by Hilbert that would decide the solvability of any Diophantine equation. Then let S be a listable set of integers—but not computable—and let P_S be the associated polynomial. By the assumption that a decision algorithm exists, there is a Turing machine program H that halts with output 1 given input integer k if the Diophantine equation

$$P_S(k, y_1, y_2, \ldots, y_n) = 0$$

has a solution, and halts with output 0 on input k if it does not have a solution.

3. But this means that the program H always halts, contradicting the assumption that the set S is listable, but not computable. Therefore, no such program H can exist.

So this is the flow of logic in Davis's argument. Unfortunately, it all hinges upon the conjecture that a listable set and a Diophantine set are one and the same. Or, put another way, it depends on being able to associate a polynomial P_S with every listable set S. Just to see how this association works for a specific case, let's look at an example due to Keith Devlin.

Let the set S consist of all integers that can be expressed as the difference of two squares. These are then the numbers that can be written as $s = a^2 - b^2$ for positive numbers a and b. So the numbers

$$1 = 1^2 - 0^2, \quad 3 = 2^2 - 1^2, \quad 4 = 2^2 - 0^2,$$

are in S, but the number 6 is not since there are no integers a and b such that $6 = a^2 - b^2$. In general, the numbers in S consist of all numbers that are not of the form $4k+2$ for some positive integer k. Thus,

$$S = \{1, 3, 4, 5, 7, 8, 9, 11, \ldots\}.$$

Consequently, if n is in S, it must have one of the forms $4k$, $4k+1$ or $4k+3$. In the first case

$$n = \left(\frac{n}{4} + 1\right)^2 - \left(\frac{n}{4} - 1\right)^2, \tag{†}$$

and in the other two cases

$$n = \left(\frac{n+1}{2}\right)^2 - \left(\frac{n-1}{2}\right)^2. \tag{‡}$$

On the other hand, every square is either a multiple of 4 or one greater than a multiple of 4, depending on whether it is the square of an even or an odd number. So the difference of two squares can never be two more than a multiple of 4, and thus numbers not in S are not a difference of two squares.

Now let's associate the polynomial

$$P_S(x, y_1, y_2) = y_1^2 - y_2^2 - x,$$

with the set S. Then it's easy to verify that a positive integer k will belong to S if and only if the equation

$$y_1^2 - y_2^2 - k = 0$$

has an integer solution (i.e., it is Diophantine).

Of course, this example works because of the special property of the set S noted above in relations (†) and (‡). What Davis needed to do is show that this kind of polynomial could be found for *any* listable set.

It's amusing to note that the prime numbers constitute a listable set, since it's possible to go through the natural numbers, one at a time, and test them for primality. So if Davis's Conjecture were true, then there should be a polynomial function whose positive values are exactly the set of prime numbers. To anticipate our story just a bit, here is such a polynomial in the 26 letters of the Latin alphabet:

$$\begin{aligned} P(a,b,\dots,z) = (k+2)\{1 &- [wz+h+j-q]^2 \\ &- [(gk+2g+k+1)(h+j)+h-z]^2 \\ &- [2n+p+q+z-e]^2 \\ &- [16(k+1)^3(k+2)(n+1)^2+1-f^2]^2 \\ &- [e^3(e+2)(a+1)^2+1-o^2]^2 \\ &- [(a^2-1)y^2+1-x^2]^2 - [16r^2y^4(a^2-1)+1-u^2]^2 \\ &- [((a+u^2(u^2-a))^2-1)(n+4dy)^2+1-(x+cu)^2]^2 \\ &- [n+l+v-y]^2 - [(a^2-1)l^2+1-m^2]^2 \\ &- [ai+k+1-l-i]^2 \\ &- [p+l(a-n-1)+b(2an+2a-n^2-2n-2)-m]^2 \\ &- [q+y(a-p-1)+s(2ap+2a-p^2-2p-2)-x]^2 \\ &- [z+pl(a-p)+t(2ap-p^2-1)-pm]^2\} \end{aligned}$$

As the letters a through z run through all the integers, the polynomial P takes on positive and negative integer values. The positive values are exactly the set of prime numbers; the negative values may or may not be the negatives of primes. Incidentally, the reader will note that the expression for P is given in terms of two factors, seeming to contradict the definition of a prime number as one that has no factors other than itself and 1. The apparent contradiction is resolved by noting that the formula produces only positive values when the factor $(k+2)$ is a prime and the second factor equals 1. This polynomial for primes was first published by James Jones, Daihachiro Sato, Hideo Wada, and Douglas Wiens in 1977. This result, incidentally, resolved a long-standing question as to whether the primes could be obtained as the values of a polynomial function.

But Davis's problem still remained: to show that there is such a Diophantine equation for *every* listable set, and conversely, that every Diophantine equation has a corresponding listable set. Enter Julia Robinson, arguably the greatest woman mathematician of the twentieth century.

A WOMAN IN MATHEMATICS

In 1975, Julia Robinson of the University of California, Berkeley, became the first woman mathematician to be elected to membership in the U.S. National Academy of Sciences. She was proposed for membership by Saunders MacLane, one of America's preeminent mathematicians. As a further indicator of her stature, Alfred Tarski and Jerzy Neyman, who were both old and not well and who didn't much care for each other, made the trip from Berkeley to Washington, DC, in order to be present at the balloting of the Academy and to explain the importance of Julia's work. In 1983, Robinson became the first woman to serve as president of the American Mathematical Society. Other honors included election to the American Academy of Arts and Sciences, an honorary doctorate from Smith College, and a prestigious MacArthur Foundation "genius" award. So who was Julia Robinson and what was her connection to Hilbert's Tenth Problem?

Julia Bowman Robinson started her mathematical career when she received her undergraduate degree from Berkeley in 1940. Discovering that her potential employers were more interested in her typing skills than her mathematical virtuosity, she enrolled in a graduate course in number theory being taught by Rafael Robinson. As often happens, the teacher fell in love with the student, and they were married the next year. Since the nepotism rules in effect at the time prevented her from teaching as a graduate assistant in the Berkeley math department, Robinson began working with famed logician Alfred Tarski for her doctorate, which she received in 1948. Her thesis showed that the arithmetic of rational numbers (i.e., ratios of integers) is all that one needs to formulate all problems of elementary number theory. It was in this same year that she began work on the Tenth Problem, which was to occupy the largest share of her professional career. In characteristic modesty, she once remarked to her sister, Constance Reid, the well-known chronicler of mathematical lives, that she did not care whether she solved the Tenth Problem herself, she just *had to know* the answer, she would not want to die *without knowing*.

In passing, it's amusing—and rather sad—to note some of the difficulties a woman mathematician faced in those years trying to secure a permanent position at Berkeley. At one point Robinson was required to submit a description of what she did each day to Berkeley's personnel office. Here's what she wrote: "Monday—tried to prove theorem, Tuesday—tried to prove theorem, Wednesday—tried to prove theorem, Thursday—tried to prove theorem, Friday—theorem false." The personnel office then let the graduate division handle Robinson's appointment. Her friend from graduate school days, Elizabeth Scott, noted that,

> Throughout her life Julia stood up for offering opportunities to all students. She also encouraged graduate students and young faculty to have more confidence in their real abilities. She felt that women and minority mathematicians especially needed this support, which she provided with spirit yet in a quiet way. She encouraged us to work together so that all women who have the ability and the desire to do mathematical research can have the opportunity to do so.

Robinson's most important contribution to the Tenth Problem was a result she obtained jointly with Hilary Putnam and Martin Davis in 1961. They proved that every listable set can be associated with an *exponential* Diophantine equation, and conversely. This solved a variant of Hilbert's problem, by replacing a normal Diophantine equation with an exponential one. The Davis-Putnam-Robinson result said that the kind of algorithm sought by Hilbert did not exist for exponential Diophantine equations. Now what exactly is an exponential Diophantine equation? These are simply equations constructed by the usual rules of polynomials from the variables—but by admitting exponentiation as one of the operations. Thus, in such equations unknown variables can appear as exponents, whereas in normal Diophantine equations the exponents can only be fixed integers. So, for instance,

$$(x+1)^{y+2} + x^3 - y^{(x+1)^x} - y^4 = 0,$$

is an exponential Diophantine equation as unknown quantities x and y appear as exponents.

Following this logjam-breaking work, in order to prove Davis's Conjecture it was sufficient to show that the set of all triples of the form (a, b, a^b) is Diophantine. In other words, in order to prove that *every* listable set is Diophantine, it was sufficient to show that *one particular* set of triples has a Diophantine representation. Julia Robinson failed to find a Diophantine representation for exponentiation. But she did manage to find a condition sufficient for the existence of such a representation. Basically, her condition involved finding a Diophantine equation whose solutions grew exponentially fast in a particular way. At this point the hunt for the Tenth moved from Berkeley to Leningrad, where a 22-year-old researcher at the Steklov Mathematical Institute pounded the final nail into the coffin.

ON GROWTH AND RABBITS

The thirteenth century Italian mathematician Leonardo of Pisa, better known by his pen name Fibonacci, published an influential book, *Liber Abaci,* which introduced the Arabic number system into Europe. One of the problems in this book involved the growth of rabbits. Fibonacci asked if you put a pair of rabbits together and they breed a new pair each month, and each new pair bred does the same, how many rabbits will there be at the end of a year?

Assuming that one month elapses before the first pair starts to produce, that there are no deaths during the year, and that each pair continues to produce regularly, it's easy to see that the number of adult pairs present month-by-month is 1, 1, 2, 3, 5, 8, 13, 21, 34, ... , a sequence generated by the rule that each term after the first two is simply the sum of the two preceding terms. This so-called *Fibonacci sequence* played the pivotal role in the final solution of Hilbert's Tenth Problem. Here's how.

What we need is to find a Diophantine equation whose solutions grow very fast, exponentially fast, in fact, as the degree of the equation becomes larger. The Fibonacci sequence is a good candidate for the creation of such an equation, as the nth term in the Fibonacci sequence is approximately equal to

$$\frac{1}{\sqrt{5}}\left[\frac{1}{2}\left(1+\sqrt{5}\right)\right]^n .$$

As n becomes large, this approximation gets better and better—and the numbers get larger and larger. Thus, by the Davis-Putnam-Robinson result for exponential Diophantine equations, to solve the Tenth Problem it was sufficient to find a normal Diophantine equation whose solutions were related to the Fibonacci numbers, since by the above formula these numbers grow exponentially fast as n gets large. This is exactly what Yuri Matyasevich, a young mathmatician in Leningrad, managed to do in 1970.

Matyasevich started with the set of ten polynomial equations:

$$
\begin{aligned}
u + w - v - 2 &= 0,\\
l - 2v - 2a - 1 &= 0,\\
l^2 - lz - z^2 - 1 &= 0,\\
g - bl^2 &= 0,\\
g^2 - gh - h^2 - 1 &= 0,\\
m - c(2h + g) - 3 &= 0,\\
m - fl - 2 &= 0,\\
x^2 - mxy + y^2 - 1 &= 0,\\
(d - 1)l + u - x - 1 &= 0,\\
x - v - (2h + g)(l - 1) &= 0.
\end{aligned}
$$

In these equations, the variables u and v are such that v is the $2u$th element in the Fibonacci sequence. So if you square each of these equations and add the resulting ten quantities together, you end up with a normal Diophantine equation whose solutions grow at a rate fast enough to satisfy the Davis-Putnam-Robinson result. This equation then solves the Tenth Problem by showing that there cannot exist an algorithm that decides the solvability of an arbitrary Diophantine equation. Finis!

The algorithmic undecidability of Hilbert's Tenth Problem is today regarded as a negative solution; it says that something does not exist. Would Hilbert himself accept this as a "solution" to the problem? Of course, there is no way to know for sure what Hilbert might or might not have accepted. But it's reasonable to suppose that he would have been surprised—and then accepted the negative result. One reason for believing this is found in a passage in his famous lecture on the 25 Problems. There he wrote,

> Occasionally it happens that we seek the solution under insufficient hypotheses or in an incorrect sense, and for this reason do not succeed. The problem then arises: to show the impossibility of the solution under the given hypotheses, or in the sense contemplated. Such proofs of impossibility were effected by the ancients, for instance when they showed that the ratio of the hypotenuse to the side of an isosceles triangle is irrational. In later mathematics, the question as to the impossibility of certain solutions plays a preeminent part, and we perceive in this way that old and difficult problems, such as the proof of the axiom of

> parallels, the squaring of the circle, or the solution of equations of the fifth degree by radicals have finally found fully satisfactory and rigorous solutions, although in another sense than that originally intended. It is probably this important fact along with other philosophical reasons that gives rise to conviction (which every mathematician shares, but which no one has as yet supported by a proof) that every definite mathematical problem must necessarily be susceptible of an exact settlement, either in the form of an actual answer to the question asked, or by the proof of the impossibility of its solution and therewith the necessary failure of all attempts.

From this it seems likely that Hilbert would have been satisfied with the Davis-Putnam-Robinson-Matyasevich solution of the Tenth Problem. But would he have been satisfied with the statement of the problem itself if he knew it would be "solved" in this way? Here we can speculate that perhaps the answer is No. Here is Yuri Matyasevich's argument for this conclusion.

In Hilbert's 1900 lecture to the Mathematical Congress, he only spoke about 10 of the 23 problems that appeared in the written version of the talk—despite the fact that he spoke for 2-1/2 hours! In particular, the Tenth Problem occupies less space in the written text than any other problem. And he gave no motivation whatsoever for including this problem on his list. So we can only guess at why Hilbert asked for solutions only in *rational integers.* We have noted above that this is equivalent to asking for an algorithm for solving Diophantine equations in nonnegative integers. But the ancients as far back as Diophantus himself were already looking for solutions in rational numbers. So why didn't Hilbert ask for a procedure to determine the existence of solutions in rational numbers, not just integers?

Matyasevich believes that the answer is that Hilbert was an optimist, believing in the existence of an algorithm for solving Diophantine equations in integers. It's a small exercise to show that such an algorithm would allow us to solve equations in rational numbers as well. So asking *explicitly* about solving Diophantine equations in integers is the same as *implicitly* asking about solving Diophantine equations in rational numbers. A positive solution of the Tenth Problem would then have immediately given a positive solution to the analogous problem about solutions in rational numbers.

But the actual result is a negative one; there is no algorithm for solving Diophantine equations in integers. What does this imply for solving these equations in rational numbers? Basically, nothing. The decision problem for the solvability of an arbitrary Diophantine equation in rational numbers involves a smaller class of Diophantine equations, and so it is at least plausible that for this narrower class of equations such a decision procedure may actually exist. At present, no one knows.

It's also of considerable interest to look at how a lot of important problems in mathematics can be reduced to special cases of the Tenth Problem, and whose solutions would have followed immediately from a positive solution to the Tenth Problem. Here are just four such problems, each of them being one of the major mathematical mountaintops that has stood the test of time and helped define entirely new fields of mathematics.

- *Fermat's Last Theorem:* As almost everyone knows by now following Andrew Wiles's well-chronicled solution of this outstanding puzzle, the Fermat problem is about the unsolvability of an infinite series of Diophantine equations

$$x^n + y^n = z^n,$$

for $n=3, 4, 5, \ldots$. So this is not quite a case of the Tenth Problem, in which Hilbert asked about solving only a single Diophantine equation rather than an infinite sequence of them. Hilbert also didn't explicitly mention this famous problem in his list of big problems for the twentieth century.

Fermat's equation is a Diophantine equation in x, y, and z for a fixed value of n, but is an exponential Diophantine equation if regarded as an equation in four unknowns, x, y, z, and n. But work on the Tenth Problem has shown us how to find a particular polynomial F with integer coefficients such that the equation

$$F(n, x, y, z, u_1, u_2, \ldots, u_m) = 0$$

has a solution in $u_1, u_2, \ldots, u_m$ if and only if n, x, y, and z are a solution of Fermat's equation. So Fermat's Last Theorem is equivalent to the statement that a particular Diophantine equation, specifically

$$F(w + 3, x + 1, y + 1, z, u_1, u_2, \ldots, u_m) = 0,$$

has no solution in nonnegative integers. Therefore, a positive solution of the Tenth Problem in its original formulation would provide a tool to prove or disprove Fermat's Last Theorem. So while the Fermat problem was not explicitly mentioned by Hilbert, it is present in his list implicitly as a very particular case of the Tenth Problem.

- *Goldbach's Conjecture:* This problem involves proving or disproving the assertion that every even number greater than 4 is the sum of two positive prime numbers. This problem was explicitly mentioned by Hilbert as part of his discussion of the Eighth Problem involving the distribution of prime numbers. Let's consider the set $\mathcal{G}$ consisting of even numbers that are greater than 2 and yet are not the sum of two primes. The elements of this set consist of counterexamples to Goldbach's Conjecture. So the Conjecture is equivalent to the claim that $\mathcal{G}$ is empty. For any particular number a it's easy to check whether it is in the set $\mathcal{G}$. So $\mathcal{G}$ is a listable set. Thus, it is possible to find a particular Diophantine equation

$$G(a, x_1, x_2, \ldots, x_m) = 0$$

that has a solution if and only if a is a counterexample to the Conjecture. Conversely, the truth of Goldbach's Conjecture is equivalent to saying that the Diophantine equation

$$G(x_0, x_1, x_2, \ldots, x_m) = 0$$

has no solution for any choice of x_0.

Again, we see that a positive solution of the Tenth Problem in its original form would have enabled us to know the truth or falsity of Goldbach's Conjecture.

- *The Riemann Hypothesis:* The main thrust of Hilbert's Eighth Problem is the famed Riemann Hypothesis, a conjecture that many feel is the single most important unsolved d problem in mathematics today. It is a statement about the location in the complex plane of the zeros of the complex-valued zeta function

$$\zeta(z) = 1 + \frac{1}{n^z} + \frac{1}{2^z} + \frac{1}{3^z} + \cdots$$

This function converges for all complex numbers z whose real part is greater than 1; the Riemann Hypothesis states that all the nontrivial zeros of this function have the form $z = \frac{1}{2} + bi$, where b is a real number. In other words, the nontrivial zeros all lie on the line in the complex plane where the real part of the complex number equals 1/2. By "nontrivial" we mean that z is not a negative even integer like $-2, -4, -6, \ldots,$ since $\zeta(z) = 0$ for all such values of z.

It turns out to be possible to construct a particular Diophantine equation

$$\mathcal{R}(x_1, x_2, \ldots, x_m) = 0$$

having no solution if and only if the Riemann Hypothesis is true. So once again we see that an outstanding mathematical problem is a particular case of Hilbert's Tenth Problem.

All the problems discussed so far—Fermat's Last Theorem, Goldbach's Conjecture, the Riemann Hypothesis—are about numbers, so the reduction of these problems to Diophantine equations is not completely unimaginable. Difficult, yes. But still plausible. What is surprising is that the reduction to the solvability of a Diophantine equation extends even to problems that are not about numbers at all. Here is one last example, the famed Four-Color Theorem that we will take up in the next chapter.

- *Four-Color Conjecture:* This is a problem about coloring maps drawn in the plane. After considerable work using advanced tools of mathematical logic, it turns out that there is a Diophantine equation

  $$C(x_1, x_2, \ldots, x_m) = 0$$

  that has no solution if and only if the Four-Color Conjecture is true.

The reduction of these mathematical mountaintops to the solvability of Diophantine equations is surprising, amusing, remarkable, and a lot of other things. But is it useful? After all, the Tenth Problem is undecidable so we have no universal method for resolving all these problems at once. Nor can we really solve any of these problems by looking at their corresponding Diophantine equation because the equations are enormously complicated, much more so than the equation given earlier for the prime numbers.

But it's often useful to recall the admonition of the great German number theorist, Leopold Kronecker, who remarked "One must always invert." So we can reverse the order of things and use the undecidability of the Tenth Problem as a challenge to invent better and better methods to solve more and more types of Diophantine equations. In this view, Wiles's proof of the Fermat Conjecture and Wolfgang Haken and Kenneth Appel's resolution of the Four-Color Conjecture can be seen as deep tools for treating particular Diophantine equations. We should perhaps try to extend these methods and techniques to other such equations. Who knows what we might find!

The Tenth Problem is about algorithms, which in modern parlance essentially means computer programs. Earlier we talked about Turing machines and computability. But so far we haven't said much about how the Tenth Problem looks when clothed in the language of computing machines and programs. So let's conclude this chapter by filling in this gap.

INFORMATION, PLEASE

Suppose we have a universal Turing machine (UTM) and consider the set of all possible programs that can be run on this machine. As we already know, every such program can be labeled by a string of 0s and 1s, so it's possible to "name" each program by its own personal ID number. Consequently, it makes sense to consider listing the programs, one after the other, and talk about the kth program on the list, where k ranges through the positive integers. Now consider the question: "If we pick a program from the list at random, what is the likelihood that it will halt when run on the UTM?" Or, equivalently, we could start with a fixed program for the UTM and ask the same question for an input string that's random. It turns out that this question is intimately tied up with the solvability of Diophantine equations, leading eventually to a remarkable result due to IBM researcher, Gregory Chaitin.

The key step in Chaitin's route to ultimate randomness was to consider not whether a Diophantine equation has *some* solution, but the sharper question of whether the equation has an infinite or a finite number of solutions. The reason for asking this more detailed question is that the answers to the original query are not logically independent for different values of k. In other words, if we know whether some solution exists or not for a particular value of k, this information can be used to infer the answer for other values of k. But if we ask whether there are an infinite number of solutions or not, the answers are logically independent for each value of k; knowledge of the finiteness or not of the solution set for one value of k gives no information at all about the answer to the same question for another value.

Following this reformulation of the basic question, Chaitin's next step was a real *tour de force.* He proceeded to construct explicitly a particular exponential Diophantine equation family specified by a single parameter k, together with over 17,000 additional variables. Let's call this equation $\chi(k, y_1, y_2, \ldots, y_{17,000+}) = 0$, using the Greek symbol χ (chi) in Chaitin's honor. From this equation we can form a very special string of binary digits in the following manner: as k successively assumes the values $k = 1, 2, 3, \ldots$, we set the kth entry in our string to 1 if Chaitin's equation $\chi = 0$ has an infinite number of solutions for that value of k, while we set the kth entry to 0 if the equation has a finite number of solutions (including no solution). As we already know, the binary string we form by this procedure represents a single real number. Chaitin labeled this number by the last letter in the Greek alphabet, Ω (Omega). And for good reason, too, as the properties of Ω show that it's about as good an approximation to "The End" as the human mind will ever make.

First of all, Chaitin showed that the quantity Ω is an uncomputable number. Furthermore, he proved that any program of finite complexity N can yield at most N of the binary digits of Ω. Consequently Ω is random, since there is no program shorter than Ω itself for producing all of its digits. Moreover, the digits of Ω are both statistically and logically independent. Finally, if we put a decimal point in front of Ω, it represents some decimal number between 0 and 1. When viewed this way, Ω can be interpreted as the probability that the UTM will halt if we present it with a randomly selected program. Or, as before, that a fixed program will halt if presented with a random input. Indeed, Chaitin constructed his equation precisely so that Ω would turn out to be this halting probability.

So while Turing considered the question of whether a given program would halt with a given input, Chaitin's extension produces the probability that a randomly chosen program will stop. As an aside, it's worth noting that the two extremes Ω equals zero or one cannot occur, since the first case would mean that no program ever halts, while the second would say that every program will halt. The trivial, but admissible, program `STOP` deals with the first case, while we leave it to the reader to construct an equally primitive program to deal with the second.

Following Matyasevich's solution of the Tenth Problem, there is a Diophantine equation $P(k, x_1, x_2, \ldots, x_m) = 0$ which has a solution if and only if the kth computer program halts. Thus, by the undecidability of the Halting Problem there is no algorithm to decide whether this equation has a solution as we let k vary over all programs.

But individual instances of the Tenth Problem are not independent. Say we have 2^N individual Diophantine equations whose solvability we want to know about. The worst case would be if this were 2^N bits of information. But it isn't. In fact, it is only $N + 1$ bits! In other words, the answers to the solvability of these equations are correlated. Why? Because if you are given the $(N + 1)$-bit number that tells you how many of these 2^N equations are solvable, then you can find which are and which are not by systematically trying all possible integer solutions until the requisite number of solvable equations are discovered.

So the Tenth Problem gives undecidability—but it doesn't give randomness. To get that start with the halting probability Ω. This number can be approximated from below as follows: the Nth approximation, Ω_N, is given by

$$\sum \frac{1}{2^p},$$

where p is the size in bits of each program of size less than or equal to N that halts in no more than N steps. In other words, you look at all programs from 1 to N bits in size. You run all of them on a UTM, giving them a maximum run time of N steps. Observe which of these programs has stopped after no more than N steps. For each such k-bit program, add $1/2^k$. This sum equals Ω_N.

Ω_N can be computed for each value of N and in the limit, $\lim_{N\to\infty} \Omega_N = \Omega$. But this convergence occurs *very* slowly. As we saw above, the bits of Ω are independent and Ω itself is random (i.e., has maximum entropy). Its bit are random mathematical facts.

To transform Ω into a Diophantine equation we use the solution to the Tenth Problem. Since the kth bit of Ω_N is computable, there is a Diophantine equation

$$P(k, N, x_1, x_2, \ldots, x_m) = 0 \qquad (\P)$$

that is solvable if and only if the kth bit of Ω_N is a 1. So the set of all N such that this equation is solvable is finite if the kth bit of Ω is a 0, and the set is infinite if the kth bit of Ω is a 1. Thus, asking for each k whether there are finitely many or infinitely many values of N for which the equation (¶) is solvable gives the kth bit of Ω. Therefore, these are independent questions, and we've found maximum undecidability. In other words, randomness.

Subsequent work by Matyasevich and James Jones on exponential Diophantine equations enables us to obtain the bits of Ω by asking whether or not the number of solutions of an infinite family of exponential Diophantine equations indexed by parameters is finite or infinite. This work implies that there exists an exponential Diophantine equation with parameters k and N that has exactly one integer solution if the kth bit of Ω_N is a 1 and has no solution if this bit is a 0. So considering N to be an unknown variable instead of a parameter leads to an exponential Diophantine equation with one parameter k that has finitely many solutions if the kth bit of Ω is 0 and has infinitely many solutions if the kth bit of Ω is a 1.

All this goes to show how ubiquitous Diophantine equations are throughout the mathematical world, and how prescient Hilbert was to focus on their solvability in his list of great challenges for twentieth-century mathematicians.

SUMMARY

Hilbert's Tenth Problem: Does there exist a single algorithm that will settle the solvability of every Diophantine equation in a finite number of steps?

Answer: No such algorithm exists.

CHAPTER TWO

THE FOUR-COLOR PROBLEM

THE SIGN OF THE FOUR

In the middle of the great southwestern desert of the United States, there is a monument marking the point where the four states of New Mexico, Arizona, Colorado, and Utah come together. If you visit this so-called "Four Corners" point, chances are you'll see tourists laying on the ground, grinning foolishly as they stretch their arms in an attempt to place one extremity in each state, their companions laughing hilariously at their contortions while they photographically record these clownish antics for posterity. Well, why not? I don't think there are many places on earth where four major political regions come together like this at a point. So the Four Corners region really is something special—and not just for tourists.

If you take a map of the 48 contiguous states of the United States, like the one shown in Figure 2.1, and try to color it so that each state has a color different from any state with which it shares a boundary, you would conclude that four colors are needed to color the four states meeting at the Four Corners point. And this is indeed the case. But you might argue that just joining at a point is not really a big enough boundary to constitute having a "shared" border. After all, a point is a zero-dimensional object, while a region in the plane is a two-dimensional one. So a true boundary should be only one dimension less than the region, a line, and not something as degenerate as a point.

Suppose then you took the Four Corners point and blew it up into a circle, thus creating a new "mini-state" taken from the territory of the four original states. The former Four Corners region would now look

Figure 2.1. A map of the contiguous 48 states of the United States.

something like the territory shown in Figure 2.2. When you come to color this new map, all of a sudden you discover that it takes only three colors instead of four. So adding the new state at the center has actually reduced the number of colors needed instead of increasing it. Now you might think about throwing away one of the states, in which case you end up with a map like the one in Color Plate I. That map has no points where any two states come together, yet now again requires four colors.

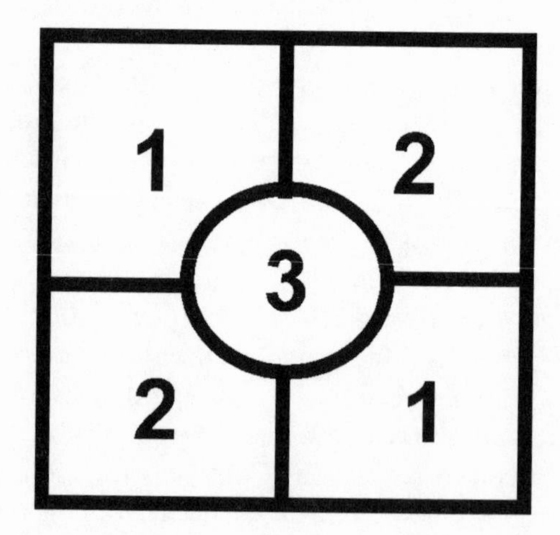

Figure 2.2. A five-region territory.

The idea of neighboring regions is, of course, the essence of the map coloring problem, and has led to much confusion about the exact nature of the Four-Color problem. You might think that if a map has no place where more than three regions come together (neighboring domains), then the map can be colored using only three colors. Wrong! In the six-region map shown in Figure 2.3, there are never more than three neighboring domains. Nevertheless, it's easy to see that this map requires four colors even though it does not contain a set of four neighboring regions. A lot of false "proofs" of the Four-Color problem have come from confusing the issue of the maximum number of neighboring domains with finding the minimum number of colors needed to color any map in the plane. For the sake of future notation, let's call the minimum number of colors needed to color any map in the plane the *chromatic number* of the plane and denote it by the Greek letter chi (χ). The maximum number of neighboring regions in the plane will be symbolized by the Greek letter nu (ν). From the map of Figure 2.2 we see that the chromatic number is always at least as big as the maximum number of neighboring regions, so $\chi \geq \nu$. The Four-Color problem for a plane is then to prove or disprove the assertion $\chi = \nu$.

Playing around with examples like this, cartographers empirically discovered that four colors always seemed to be enough to color any kind of map that could be drawn on a piece of paper (a plane or surface of a sphere), assuming the map didn't contain degenerate points like the Four Corners or regions split into more than one piece, like the Upper Peninsula of Michigan, which is also shown in Figure 2.1. On the other hand, it's easy to find examples like Color Plate I that require four colors. But is it *always* enough? Obviously, this four-color *conjecture* was crying out for mathematical attention, which it finally received in the middle of the nineteenth century. But before turning to that story, let's take a few more pages to make the nature of the Four-Color Problem more precise, as well as to show how things can change dramatically if we leave the cozy confines of a plane and look at maps drawn on more complicated surfaces.

First, let's agree that we will only consider maps having no points like the Four Corners or disconnected regions like the two parts of Michigan. Without the first condition, we could construct maps requiring as many colors as we wish. For instance, if you want to require the use of 17 colors, simply draw a circle and divide it into 17 sectors like pieces of pie. Each sector meets at the point at the center of the circle, hence each sector would require a different color. So we want to rule out this kind

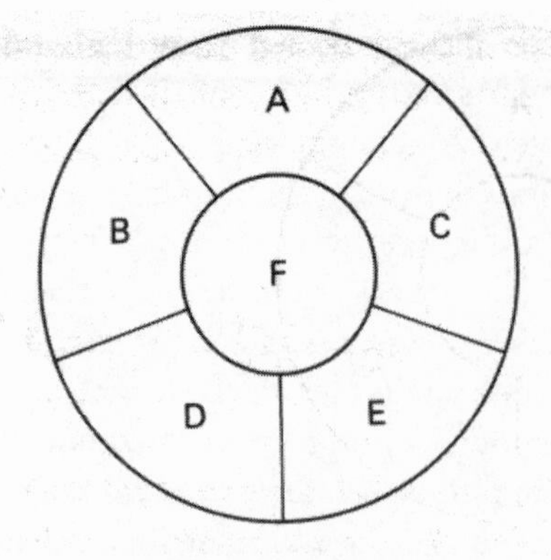

Figure 2.3. A map with at most three neighboring domains requiring four colors.

of degeneracy. Similarly, if a region is separated into two disjoint pieces, where each piece must have the same color, it's not too hard to construct maps requiring more than four colors. Figure 2.4 shows an example of such a map requiring 12 colors for its 12 domains, each of which consists of two disjoint pieces.

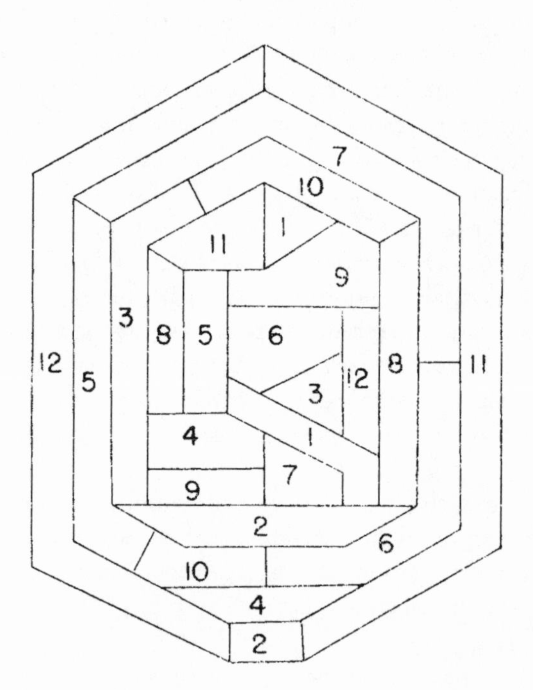

Figure 2.4. A map requiring 12 colors.

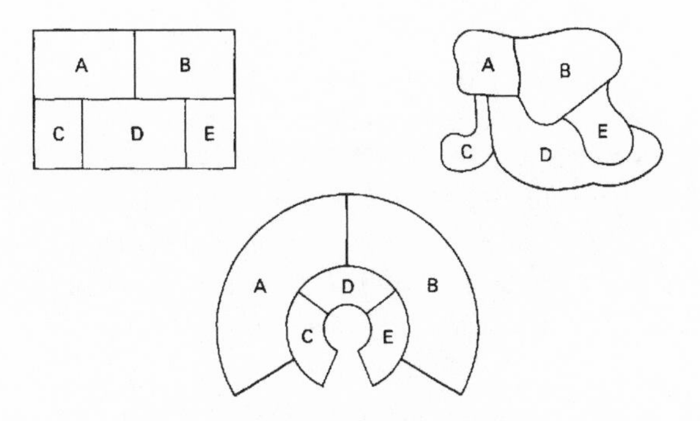

Figure 2.5. Three topologically equivalent maps.

Let's now consider another fundamental aspect of the Four-Color problem. Look at the three maps shown in Figure 2.5, and concentrate only on the relative positions of the regions. It's easy to see that the mutual boundary relationships are exactly the same for each map. So the coloring problem is the same for all three maps. In fact, a coloring of any of the three maps can immediately be transferred to the other two. So it is completely irrelevant to the coloring problem whether a boundary line, originally curved, has been straightened, or is later curved again. It also doesn't matter if we shrink or increase the size of a region. It's the *relative* positions of the boundaries that matters. And as long as the changes we make to the regions don't affect these relative positions of the boundaries, then the maps are completely equivalent insofar as how many colors are needed to color them. Relationships like the relative position of boundaries that remain unchanged when we stretch, shrink, twist of in other ways deform a shape, are called *topological,* and shapes like the three maps of Figure 2.5 are called *topologically equivalent.*

It should be clear that the map coloring problem depends essentially on the kind of surface on which the map is drawn. For instance, drawing a map on the surface of a bicycle tire, technically termed a *torus,* with its hole in the center, changes dramatically how many colors are needed to ensure that no two neighboring domains receive the same color. The hole matters. Color Plate II shows an example of a map requiring seven colors on the torus, and the map of a double torus, which requires eight colors. It can be proved for the torus we have $\chi = \nu = 7$ and for the double torus $\chi = \nu = 8$. Strangely enough, the map coloring problem was completely solved for these surfaces, which are much more complicated than the plane or sphere, topologically speaking, much sooner than for maps drawn on these simpler surfaces.

Looking at the torus, it's easy to see that it is formed by simply attaching a handle to a sphere. Topologists have shown that most surfaces of interest have this structure: they are spheres with handles, the plane being a surface topologically equivalent to a sphere with no handles. Technically speaking, the number of handles on the surface is called its *genus*, denoted by p. It has been proved by Ringels and Young in 1968 that the chromatic number of a surface of genus p, the number of colors needed to color any map on the surface, equals Heawood's number

$$H = \frac{1}{2}\left(7 + \sqrt{1 + 48p}\right),$$

for all p greater than 0. So for the torus, which has $p = 1$, the Heawood number is 7, while the double torus, having $p = 2$ has chromatic number 8 (obtained by taking the integer part of H). The problem with this formula is that its proof does not hold for the case of the plane ($p = 0$), even though putting $p = 0$ into the formula gives the right number, four. So, again, we have an example of how the map coloring problem was solved for cases of surfaces far more complicated than the one we're really interested in—the plane. Here's another example of this odd phenomenon.

Consider the surface obtained by taking a strip of paper, giving it a half twist, and then gluing the two ends together. The end result of these operations is something called a Möbius band, a surface that has only one side and one edge. Color Plate III shows both sides of a map on a Möbius band that requires six colors. Unlike the plane or the sphere, the Möbius band is a nonorientable surface, which means that it is not possible to define the notion of right- and left-handedness in a consistent fashion over the entire surface, as shown in Figure 2.6. Again, it's interesting to observe that the map coloring question was completely resolved for these nonorientable surfaces long before the final solution was obtained for orientable surfaces like the plane and the sphere.

Finally, let's note that insofar as map coloring goes, it makes no difference whether the map is drawn in the plane or on the surface of a sphere. If we start with a map on a sphere we can deform it to an equivalent map on the plane by puncturing the sphere in the middle of one of the regions and pulling the entire map out flat, so that the punctured region becomes one that surrounds the rest of the map. Conversely, if we are given a map on the plane we can regard the region surrounding the map as an extra country, and then fold the entire map into the shape of a sphere. This brings the added region together to form an "enclosed" region just like

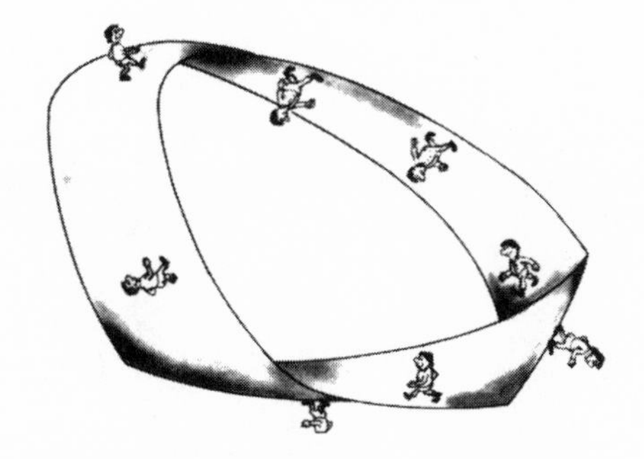

Figure 2.6. Nonorientability of a Möbius band.

all the others. The end result is that if every planar map can be colored with four colors, so can every map on the sphere, and vice versa.

All these preliminaries now in hand, finally we can formally state the

Four-Color Conjecture (4CC). No more than four colors are needed to color any map drawn in the plane (or on the surface of a sphere).

As we tell the story of how the Four-Color problem became the Four-Color Theorem, it will be useful to translate the statement of the problem from regions in a plane to the properties of particular networks of points and lines. Let's show how any map in the plane can be drawn as such a network, and then rephrase the Four-Color problem in the language of such networks.

Suppose we are given a map. Within each region of the map place a single point (for example, at the capital city of the region or its main transportation center). Each such point will be a node of the network. Now draw a link between two nodes if and only if the respective map regions for the nodes share a common boundary. In this event, the link joining the nodes must lie entirely within the two regions, crossing over the common boundary (in other words, the link cannot cross the territory of a third region). Such a neighborhood network is sometimes called the *dual graph,* or just the *graph,* of the map. An example of such a neighborhood network is shown in Figure 2.7, where the original map is on the left and its neighborhood network on the right. (In this example it is possible to join each of the nodes with straight lines, but this isn't always the case and curved connections are allowed. But this doesn't change any of the topological character of the problem, since straightness and curving are not distinguishable, topologically speaking.)

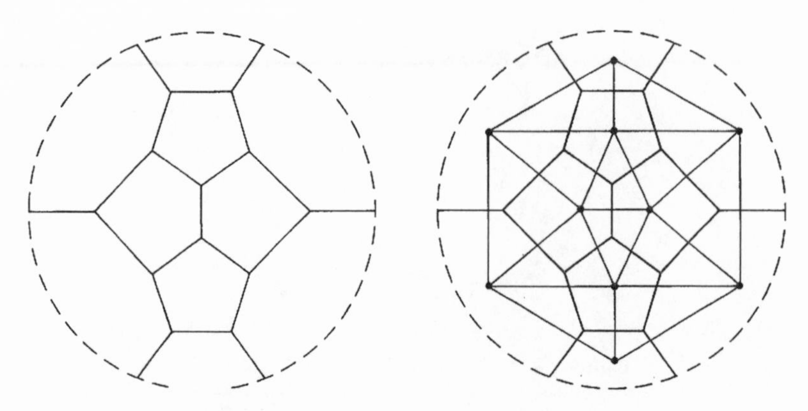

Figure 2.7. A neighborhood network for a map.

A crucial property of neighborhood networks is Euler's formula, discovered centuries ago by the famed Swiss mathematician, Leonhard Euler. It is an expression linking the number of nodes, links and regions in a map. If we denote the nodes (vertices) by V, the links (edges) by E, and the regions (faces) by F, Euler's result says that for planar maps we must have

$$V - E + F = 1. \qquad (*)$$

We'll use this formula a number of times in our subsequent discussions of the proof of the 4CP.

From the neighborhood network we see immediately the topological structure of the map it represents. The problem of coloring the map can now be restated in terms of coloring the network. The nodes must be colored such that any two nodes that are connected by a link must have different colors. If every such network can be colored by no more than four colors, then the Four-Color problem is solved in the affirmative. On the other hand, if even one network exists requiring five colors, then the conjecture that four colors suffices is false.

Notice that the way the neighborhood network was defined requires that no two paths in the network can cross. For graphs (in the sense of graph theory, not curves representing equations drawn on "graph paper"), this restriction is absent. In graph-theoretic terms, a neighborhood network is what is termed a *planar graph.* We will use the terms interchangably in what follows.

Now a bit of history to set the stage for consideration of the controversial, computer-assisted proof of this conjecture.

MAPS OF MANY COLORS

The Origin of the Four-Color Problem

Francis Guthrie was a Renaissance man—botanist, lawyer, and mathematician—in whose honor the South African flora *Guthriea capensis* was named. He was also a professor of mathematics at Cape Town and general secretary of the South African Philosophical Society. One day in 1852, Guthrie was coloring a map of the various counties in England. As he went about this task, it occurred to him that it should be possible to color not only this map, but *any* map in which the regions have no point boundaries, with not more than four colors. Putting on his mathematical hat, Guthrie struggled for some time to find a proof of this fact. Finally, in October he communicated this problem to his younger brother Frederick, who was studying under the well-known mathematician Augustus de Morgan at University College in London. On October 23, Frederick Guthrie communicated the problem to de Morgan.

De Morgan, one of the founders of the London Mathematical Society and its first president, was fascinated by the problem, and on the same day sent a letter to Sir William Rowan Hamilton in Dublin, in which he explained the problem. That letter still resides in the archives at Trinity College, Dublin, A transcript from de Morgan's original hand-written note is reproduced here in Figure 2.8. This letter appears to be the first written correspondence about the 4CP other than Guthrie's original letter to his brother, of course. (Note: Near the end of the letter de Morgan mentions the "Sphynx," in reference to the Sphinx of ancient Greek mythology, who plunged to her own death when Oedipus solved a difficult riddle she had posed to him.)

Hamilton, in contrast to de Morgan, did not find the question at all captivating, and four days later replied that he would not soon be working on de Morgan's "quaternion of colors" problem. Unfazed by Hamilton's dash of cold water, de Morgan continued to discuss the problem with colleagues, so much so, it seems, that many considered him the originator of the problem. But the first time the mathematical world, at large, learned of the true originator of the 4CP was in the article "Note on the Colouring of Maps" published by Frederick Guthrie in 1880 in the *Proceedings of the Royal Society of Edinburgh.* To de Morgan's credit, Guthrie states in this article that de Morgan had always acknowledged by whom he had first been informed of this problem. So unlike many statements and solutions of later famous mathematical problems, there is no priority dispute about the 4CP (or its solution, for that matter).

My dear Hamilton

⋮

A student of mine asked me to day to give him a reason for a fact which I did not know was a fact, and do not yet. He says, that if a figure be any how divided and the compartments differently coloured so that figures with any portion of common boundary line are differently coloured – four colours may be wanted but not more. The following is his care in which four are wanted.

A B C D are names of colours

Query cannot a necessity for five or more be invented. As far as I see at this moment, if four ultimate compartments have each boundary line in common with one of the others, three of them inclose the fourth, and prevent any fifth from connexion with it. If this be true, four colours will colour any possible map without any necessity for colour meeting colour except at a point.

Now, it does seem that drawing three compartments with common boundary A B C two and two – you cannot make a fourth take boundary from all, except inclosing one – But it is tricky work and I am not sure of all convolutions – What do you say? And has it, if true been noticed? My pupil says he guessed it in colouring a map of England. The more I think of it, the more evident it seems. If you retort with some very simple case which makes me out a stupid animal, I think I must do as the Sphynx[b] did. If this rule be true the following proposition of logic follows:–

If A B C D be four names of which any two might be confounded by breaking down some wall of definition, then some one of the names must be a species of some name which includes nothing external to the other three.

Yours truly

A De Morgan

Oct 23/52

Figure 2.8. De Morgan's letter to Hamilton explaining the Four-Color problem.

It's an interesting sidelight to note that in a review of a book by de Morgan in 1860, he says "Now, it must have been always known to map-colourers that *four* different colours are enough." Since there is not a shred of evidence in support of the belief that cartographers knew any such thing, one can only presume that here de Morgan was trying to smoke out some response from the scientific community about the problem. Even a hundred years later, Kenneth O. May, a well-known historian of mathematics, examined the huge atlas collection in the Library of Congress. He found no evidence that map colorers have tried to employ minimal colorings in the production of maps. So while some map colorers may well have suspected such a minimality result, they certainly never published their speculations nor did they make any effort to employ them in the actual coloring of maps.

Despite these promotional efforts by de Morgan, the 4CP never really caught on in the mathematical world until June 13, 1878, when the English mathematician Arthur Cayley posed the question of whether anyone knew a proof of the 4CP before the assembled members of the London Mathematical Society. No such proof was offered at the time, and Cayley's query was later published in the Society's Proceedings. This appears to be the first statement of the problem in print, as opposed to the earlier private correspondences. As Holmes would have said to Watson, at this stage "the game's afoot."

Development of the Four-Color Conjecture (4CC). Despite the fact that he was unable to prove the 4CC, de Morgan did manage to obtain one very important result. He showed that there can be no map where each of five countries borders on four others. Considering the map as a network, what this amounts to is that it's impossible to draw a network with five nodes so that each node is connected to four others without having at least one of the links crossing one of the links already drawn. An example of this is shown in Figure 2.9, where no matter how you try to join up the nodes, you will be left with two nodes (A and E) that cannot be joined without crossing one of the links already drawn. But figures are only to stimulate a sluggish imagination. Here is a simple—yet rigorous—proof of this fact.

Suppose you could draw a network with five nodes such that each node is connected to the other four without any crossings of links, regarding the area surrounding the network as an additional "face" of the network. With this convention, each link of the network will separate two faces. Moreover, as there is now one extra face, Euler's formula now becomes

$$V - E + F = 2.$$

For this network, $V = 5$. And since every node is joined by a link to every other, $E = 10$. So we must then have $F = 7$ if Euler's formula is to hold. Now let's perform another calculation.

Each face is surrounded by at least three links; otherwise, it could not be a region in the plane. But from the earlier argument the number of faces is seven. So there must be at least $3 \times 7 = 21$ links. But if you

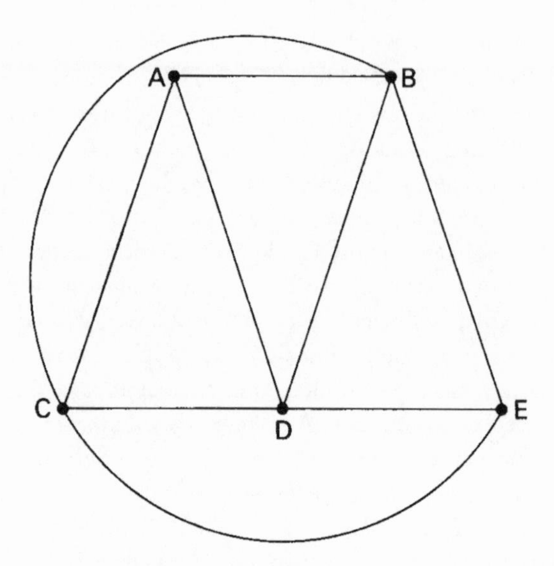

Figure 2.9. De Morgan's Theorem.

count links this way (by faces), each link will be counted twice, since it separates two faces. So it must be the case that the actual number of links is $21/2 = 10\frac{1}{2}$ links, which since there are no half links means that there are at least 11 links. But, as noted above, $E = 10$. This contradiction shows that our original assumption is false; hence, it is not possible to draw a network with five nodes so that each node is connected to the other four without any crossings of the links. This concludes the proof of de Morgan's theorem for maps, which sets the stage for the first major result on the 4CP: The Five-Color Theorem.

Cayley presented the 4CP to the London Mathematical Society in 1878. One year later one of its members, a lawyer named Alfred Bray Kempe, published a paper claiming to prove the 4CC. Eleven years later, Percy John Heawood noted a major error in Kempe's proof, but was able to salvage enough of it to prove that five colors are always sufficient to color a map. Heawood's proof of this Five-Color Theorem is important, as it makes use of Kempe's idea of *reduction*, which forms one of the pillars upon which the ultimate proof of the 4CC rests. So let's follow Keith Devlin's account of Heawood's proof.

Consider maps drawn on a sphere. As we just saw, Euler's formula in this case is

$$V - E + F = 2.$$

The idea of the proof is to begin with an arbitrary map drawn on a sphere, and gradually modify it by a process of merging two or more adjacent countries into one, so that eventually a map is obtained having at most five regions. Then by the result we just proved, we can color this resulting map with not more than five colors. Provided the steps in the modification process do not reduce the number of colors needed to color the map, this will prove that five colors suffice for the original map. Therefore, the essence of the matter is to describe the various modification steps used to reduce the map to a simpler one without reducing the number of colors needed to color the resulting map. Kempe gave six different reduction processes, each of which is applicable to a different situation depending on the specific configurations of regions on the map. These reduction processes are depicted in Figure 2.10, where the original configuration is on the left and the reduced version is on the right. Here is a description of how each of the six reduction processes works:

i. If one region is contained entirely within the other (like the country of Lesotho, which is surrounded by South Africa), then the inner region may be merged with the surrounding one. In this case, the old map can be colored with the same number of colors as the new map by assigning the inner region a color different from the one used to color the entire merged region in the reduced map.
ii. If there is a node at which more than three regions touch, then it must necessarily be the case that one pair of these regions will not have a common border anywhere on the map. These two regions are then merged into one. Given any coloring of the reduced map, the original map can be colored using the same number of colors by assigning the same color to the two regions that were merged, and coloring the rest of the map the same in both cases. This reduction process enables us to modify the original map so that only three regions touch at any point. (Example: this reduction process can be used to merge Colorado and Arizona at the Four Corners point into a single region.)
iii. If there is a region bordering on just two others, then that region can be removed by merging it with one of these two. If the reduced map can be colored using at least three colors, then the original map can be colored using the same colors by coloring the merged central region differently from the two surrounding regions.
iv. Any region having three neighbors can be removed by merging it with one of its neighbors. As in case (iii) if the reduced map can be colored using at least four colors then the original map can be colored using the same colors.

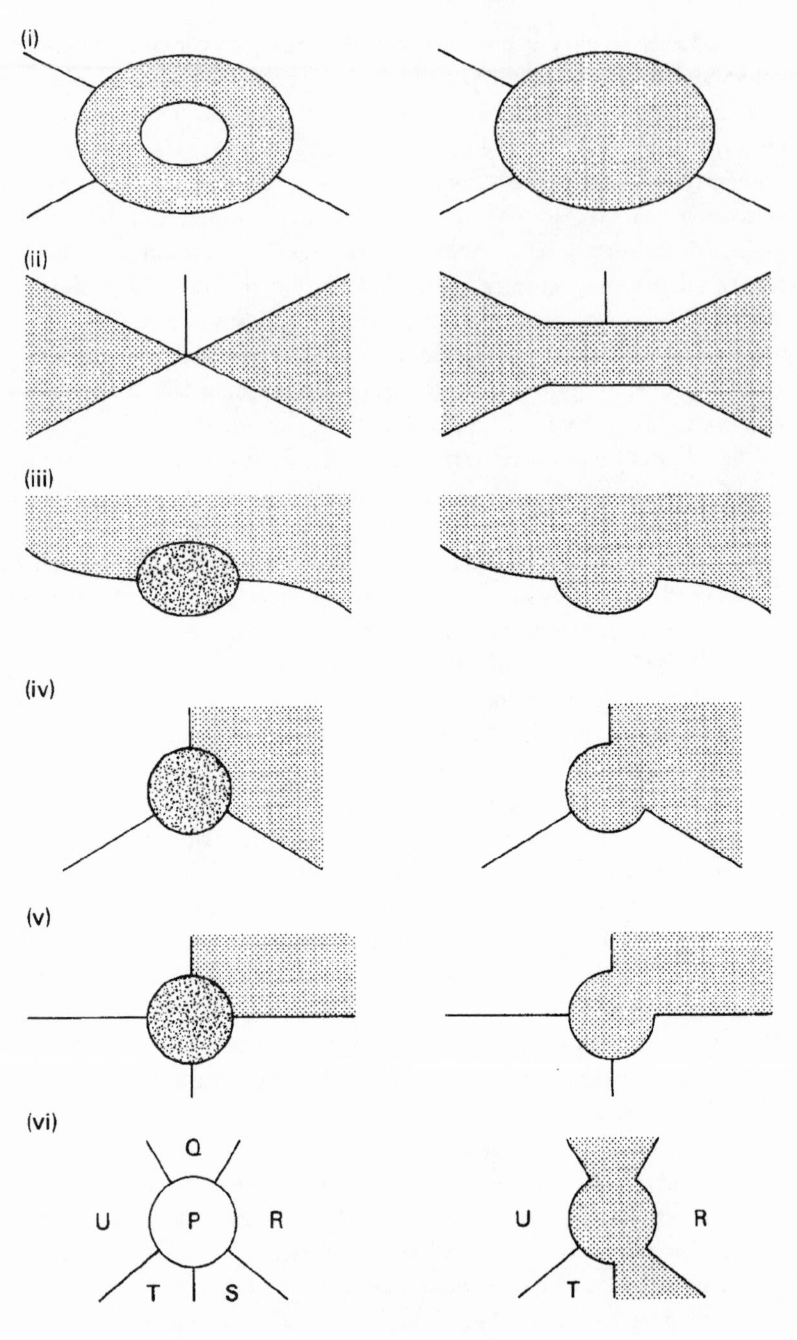

Figure 2.10. The six Kempe reduction processes for a map.

v. Any region having four neighbors can be merged with one of its neighbors, and this will not involve any change of color if five colors are available.

Applying these five reduction processes repeatedly, we will end up with a map in which no region is surrounded by another, in which each node lies on exactly three links, and all of whose regions have at least five links. In fact, it can be proved using Euler's formula that at least one region will have exactly five links.

Now consider such a five-edged region *P*, with neighbors *Q*, *R*, *S*, *T* and *U* as in part (vi) of Figure 2.10. Then we have the last reduction process

vi. One pair of neighbors of *P* do not touch. In the diagram of Figure 2.10 (vi) these are regions *Q* and *S*. Merge the three regions *P*, *Q*, *S*. If the reduced map can be colored with five colors, so can the original. In the reduced map *Q* and *S* have the same color, so there are four colors surrounding *P*. This leaves one color with which to color *P*.

The above account completes the Kempe reduction processes. Since each step reduces the number of regions on the map, by applying these processes repeatedly you eventually end up with a map having five or fewer regions. Since any such reduced map can be colored using not more than five colors, so can the original map. Note, however, that the argument used in parts (v) and (vi) don't work when only four colors are available. Thus we have the

Five-Color Map Theorem. Every map on the plane or sphere can be colored using not more than five colors.

It's worth taking a bit more time examining the structure of Kempe's failed proof, since the actual solution by Haken and Appel was basically an extension of the strategy originally employed by Kempe.

Kempe's approach was to assume the existence of a map requiring five colors, then to obtain a contradiction thereby showing that only four colors are actually needed. He began by defining the idea of a *normal* map. This is one in which no region is entirely surrounded by another and no more than three regions meet at a point. Starting from a map that requires five colors, you can always obtain a normal map by employing reduction processes (i) and (ii). So there is no loss of generality in assuming we are working with a normal map. Of course, there may be many normal maps requiring five colors, with different numbers of regions. Among this set of normal maps there will be at least one having the smallest number of regions. Kempe tried to obtain his contradiction by working with such a minimal normal map requiring five colors.

It's important to note that the point of using a minimal map is that any normal map with fewer regions may be colored using four colors. So if you can find a reduction process that will reduce the size of the minimal map by even one region, without altering the requirement for five colors, then you will have obtained a contradiction since the reduced map cannot at the same time be colored with four colors and not colored with fewer than five colors.

In his "proof," Kempe correctly proved that in any normal map there has to be some region with at most five neighbors, i.e., there is no map in which every country has six or more borders. Stated differently, at least one of the configurations shown in Figure 2.10 (iii), (iv), (v), and (vi) has to turn up somewhere in the map. He then argued (wrongly) that if a minimal normal map requiring five colors has a country with at most five neighbors, then it could be reduced to a normal map with fewer countries that still required five colors. This would have given him the contradiction just noted. Where Kempe went astray was in the case of a country with five neighbors (part (vi) of the reduction processes). To describe the flaw in Kempe's argument, we need the notion of an unavoidable set of configurations.

Another way of expressing the fact that no normal map can exist in which every country has six or more borders is to say that the set of configurations (arrangements of countries) consisting of a country with two borders, a country with three borders, a country with four borders, and a country with five borders is *unavoidable*, in the sense that every normal map must contain at least one of these four configurations. An *unavoidable set of configurations* is a collection of map configurations such that any minimal normal map requiring five colors must contain at least one of them. Kempe's attack on the 4CC was essentially an attempt to find an unavoidable set of reducible configurations. His unavoidable set consisted of the configurations shown in parts (iii)-(vi) in Figure 2.10. Kempe's proof of the 4CC failed because his proof of reducibility didn't work for the configuration in part (vi).

Despite its failure, Kempe's argument involved two of the key ideas that would eventually be used in the Haken-Appel solution to the 4CC. The first is the notion of an unavoidable set, while the second is the idea

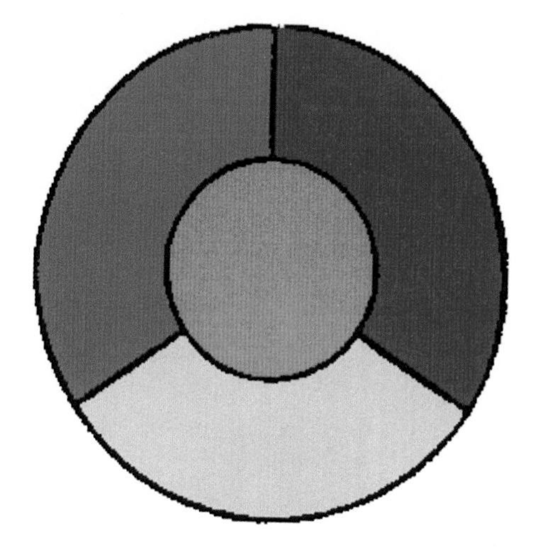

Color Plate I. A map requiring four colors.

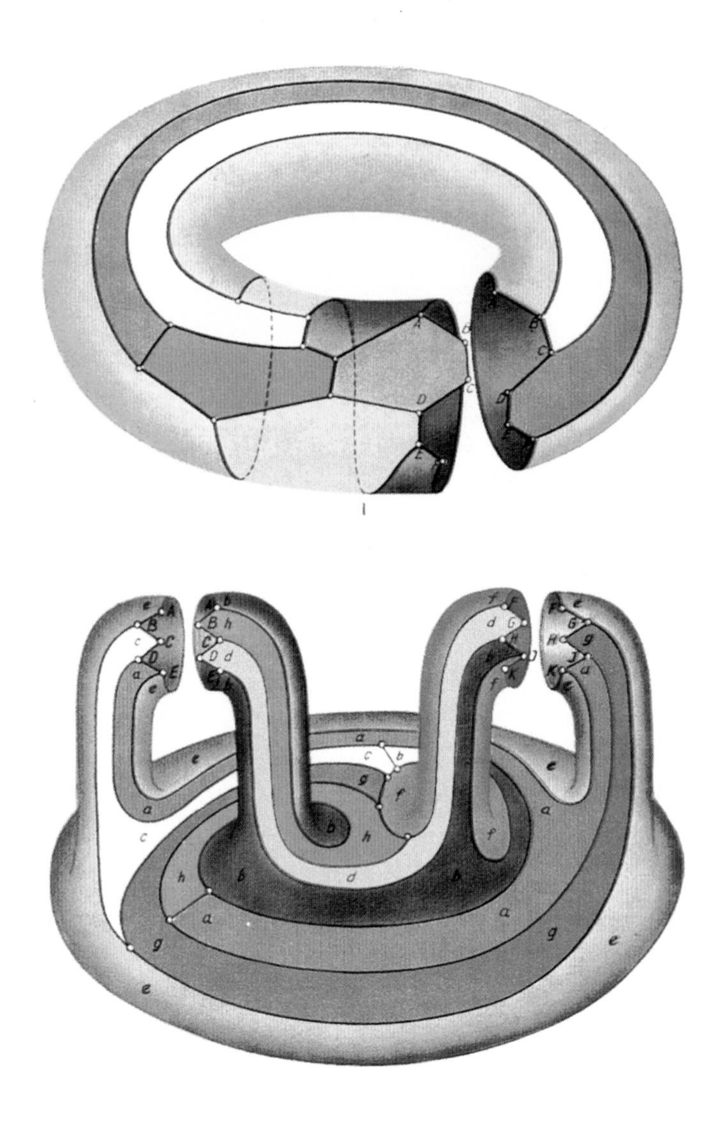

Color Plate II. Maps on the torus and the double torus.

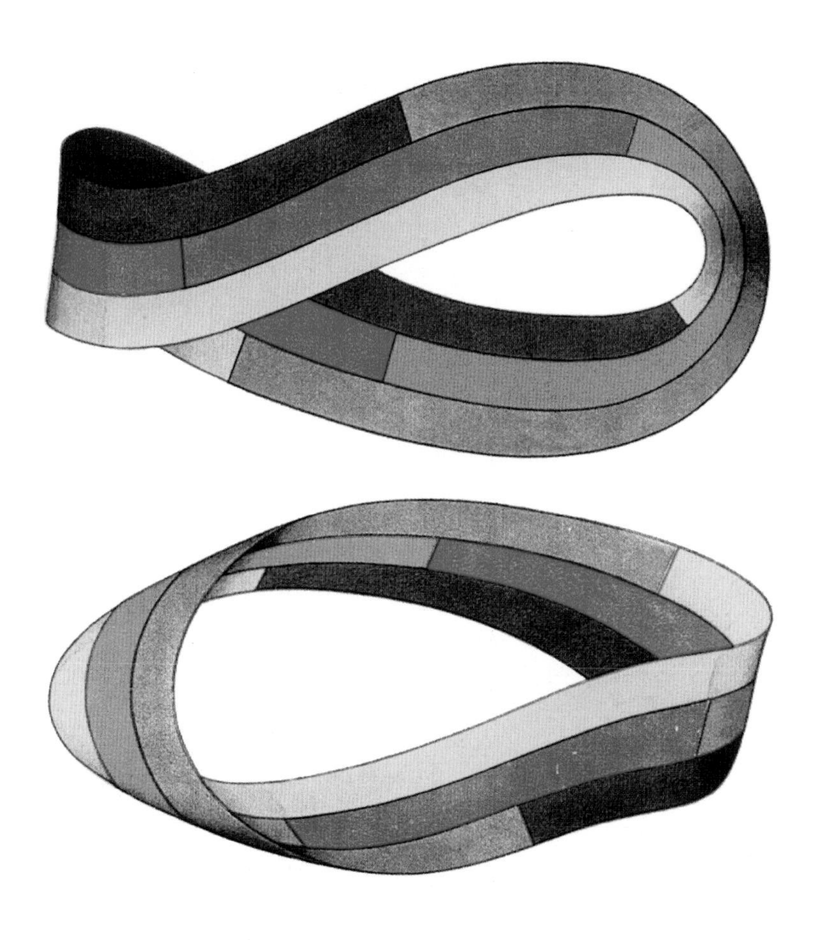

Color Plate III. A map on the M\"obius band requiring six colors.

of reducibility. Provided you can demonstrate that every configuration in an unavoidable set is reducible, you can prove the Four-Color Theorem. Intuitively, the reason for this is that a configuration is reducible if one can show that it cannot possibly appear in a minimal normal map colorable with five colors. The Haken and Appel proof succeeded because they were able to make a detailed analysis of the flawed case (vi) in Kempe's argument, which required the discovery of a different unavoidable set. But instead of having only four configurations like Kempe's unavoidable set, the Haken-Appel unavoidable set had over 1,400 configurations! We'll come later to how such a set was discovered. But now back to the history book.

Earlier, we saw Heawood's formula for the solution of the 4CP for maps drawn on all surfaces with the exception of the plane and the sphere. After noting the gap in Kempe's argument in 1890, Heawood spent the next 60 years working on the problem. His development of the formula given earlier in the chapter employed the same kind of argument used to prove the Five-Color Map Theorem for maps drawn on a sphere. This involved making essential use of the *Euler characteristic* of a surface. On any surface, the quantity $V - E + F$ will be unchanged for any map drawn on that surface. For the sphere, the value of this quantity is 2. Of course, for different surfaces the Euler characteristic will have different values. So, for instance, on the sphere we have already seen that the value is 2, while for the torus the value is 0. But it remains fixed no matter how the surface is topologically twisted, bent, or deformed. Thus, it is a *topological invariant* of the surface. Note, however, that while two topologically equivalent surfaces have the same value of the Euler characteristic, the converse is not true. Two surfaces may have the same Euler characteristic but not be topologically deformable, one to the other. For example, the torus and the Klein bottle, a surface having no edges and only one side, both have Euler characteristic 0. Yet, it is not possible to deform one to the other without cutting one of the surfaces. So they are not topologically equivalent.

The last major development in the "middle ages" of the 4CP was made by the celebrated Harvard mathematician, George Birkhoff. In 1913 he improved on Kempe's reduction techniques by introducing the notion of *reducible rings,* which are sets of nodes in a network graph that can be arranged so as to form a closed chain. By this innovation he was able to show that certain configurations larger than Kempe's were reducible. One of Birkhoff's students, Philip Franklin, used these ideas to show that every map of 25 regions or less could be colored with four colors.

Various other researchers managed to up this figure to 40 regions by 1970, at which point the "modern age" of the 4CP began.

Solution of the Four-Color Problem. The set of all configurations that had been proved reducible by 1970 wasn't even close to forming an unavoidable set. A number of unavoidable sets had been generated, but none seemed as if they would be reducible. So the situation appeared to be that either you could have reducibility or you could have unavoidability—but not both. Enter Heinrich Heesch.

After the rise of National Socialism in Germany in the 1930s, Heesch was forced to leave the university, spending the next 20 years working as an independent scholar, advisor to industry and teacher of music. Remarkably, in 1935 Heesch published a counterexample to part of the 18th Problem on the famous list given by Hilbert at the International Congress of Mathematicians in 1900. Only in 1955 did Heesch again resume teaching at the Institute of Technology in Hannover. At the end of the 1940s Heesch became the first mathematician (after Kempe) to publicly state that he thought the 4CC could be proved by finding an unavoidable set of reducible configurations. In 1950 he conjectured that the configurations in such a set would be limited in size, and that the set would have about ten thousand members. While this limit proved wildly overgenerous, it did point out that the problem would most likely only be solved using powerful computers having the ability to handle large sets of configurations.

Heesch's student Karl Dürre wrote a program to prove reducibility. This work made use of the fact that normal maps, when expressed as neighborhood networks, have each node joining exactly three links. So all of the faces (regions) in the network are triangles, and the network itself is called a *triangulation* of the map. In the language of neighborhood networks, a configuration is a part of a triangulation consisting of a set of nodes plus all of the links joining them. The nodes corresponding to the ring of countries bounding the countries of the configuration, along with the links joining them form what is called the *ring* of the configuration. To illustrate, the four neighbors A, B, C, and D of the single country E in Figure 2.11 constitute the ring of the configuration consisting of the single country E. During his work on reducibility procedures, Heesch discovered a number of "reduction obstacles"—conditions preventing reducibility of configurations—that were incorporated into the computer programs to greatly improve their efficiency. So with Heesch's work on reducibility well-developed by the late 1960s, all of the ideas on reducibility needed to solve the 4CC were in hand.

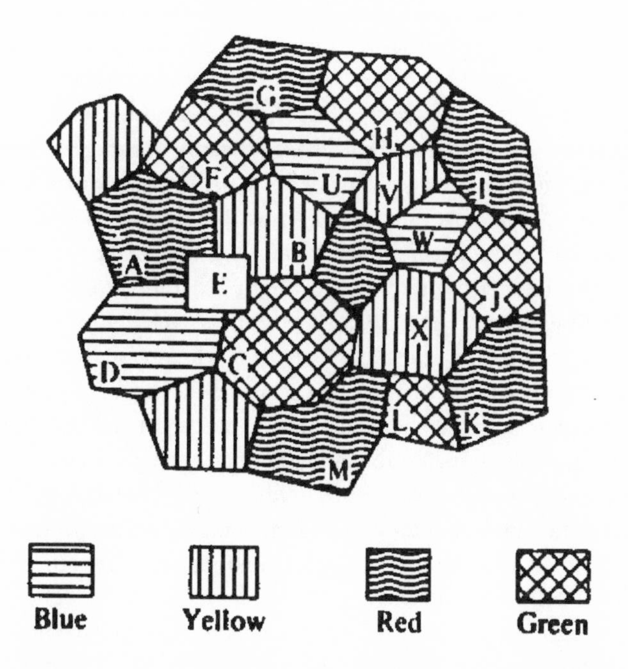

Figure 2.11. The ring of the configuration consisting of the single country *E*.

Unfortunately, progress on finding unavoidable sets of configurations had not been nearly so rapid as for reducibility. Heesch then developed the outline of a method now called *discharging* for systematically finding sets of unavoidable configurations that opened the way to the final solution. Here is how the method works.

By Kempe's work, a triangulation representing a minimal normal map colorable with five colors cannot have any nodes with fewer than five neighbors. So assign the number $6 - k$ to every node of degree k. That is, to every node having k neighbors. The sum of the assigned numbers (called "charges") is exactly 12, a fact depending on the planarity of the network and that the network is a triangulation. The particular value 12 is not very important. But what is important is that for every planar triangulation the sum of the charges is positive. This numbering scheme also shows that nodes of degree greater than six are assigned negative charges.

Now suppose the charges in a triangulation are moved around (in fractions even) so that the overall charge in the network remains unchanged. In particular, positive charge is moved from some of the positively charged (degree five) nodes to some of the negatively charged (degree greater than six) nodes by these operations, so that nodes originally having positive charge may change. This means that some degree-5 nodes may lose all their charge (become "discharged"), while some nodes of degree greater than 6 may end up with a positive charge (become "charged"). The exact final arrangement of charges will depend on the precise redistribution procedure employed. But because it's possible to detemine the layout of small parts of the network without knowing the entire net, with a specific discharging procedure it is possible to generate a finite list of all the configurations that will end up with a net positive charge.

Now, since the total charge on the network is positive, there will always be some nodes with positive charge. But as all possible receivers of positive charge are included in the finite list of configurations that has been generated from the discharging procedure, every network must contain at least one of these configurations. Put differently, the list of configurations that is generated will form an unavoidable set.

Figure 2.12 shows a simple example of the generation of an unavoidable set with the following discharging procedure. First transfer $\frac{1}{5}$ of a unit of charge from every degree-5 node to each neighboring node having degree 7 or more. Then the resulting unavoidable set consists of just the two configurations shown in the figure. To see how these come about, note that a degree-5 node can end up with positive charge only if it has at least one neighbor of degree 5 (left side of the figure) or degree 6 (right side of the figure). A degree-6 node starts with no charge, and does not receive any under this discharging method. A degree-7 node can become positive only if it has at least six degree-5 neighbors. If this happens, since every face is a triangle, two of these neighbors are joined by a link (which means that the left side of the figure applies to that pair of neighbors). A node of degree 8 or higher cannot become positive even if all its neighbors have degree 5. Moving $\frac{1}{5}$ of a unit simply is not enough. Thus the two configurations shown in Figure 2.12 form an unavoidable set. In other words, at least one of these two configurations will always be found. In the part (i) of the figure, the configuration is produced by two degree-5 nodes joined together, while the configuration in part (ii) is constructed from a degree-5 node joined to a degree-6 node. These two node pairs are shown as black circles connected by a heavy link. The remainder of each configuration is then determined by the degrees of the nodes in the producing pairs and the fact that the network is triangular. The outer nodes (the white circles) have no restrictions on their degrees.

The point of this "discharging" of positive nodes is to find a systematic procedure for how to move charge so that every node of positive charge

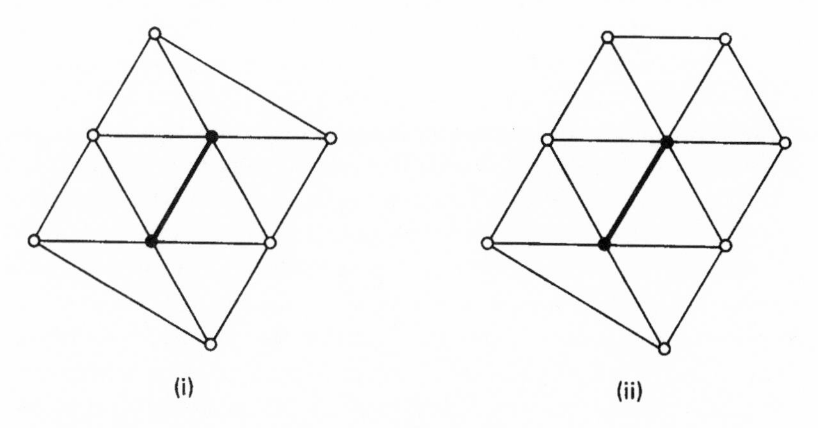

Figure 2.12. An unavoidable set, consisting of the configurations (i) and (ii).

in the resulting distribution must belong to a reducible configuration. In that case, since every triangulation must have nodes of positive charge, the configurations singled out by this discharging procedure must form an unavoidable set and the 4CC will be proved. Of course, if not all of the configurations resulting from this discharging procedure are reducible, there will be no help. In fact, Kempe's original unavoidable set, the one that caused his proof of the 4CC to fall apart, corresponds to the procedure of moving no charge at all!

Many know the work of the French lyricist Paul Valéry. But what most don't know is that in 1902 Valéry anticipated some of the work on the 4CP by Birkhoff, Franklin and others in notes found in his diary by French literature expert, Jean Mayer. Mayer, himself developed an interest in the problem, and spent time working on how to improve discharging procedures. This work led to a collaboration with Appel and Haken, and was eventually incorporated into the computer program used to actually prove the 4CC. But we're getting ahead of the story. So without further ado, let's now give a full account of the magnificent accomplishment by Haken and Appel that finally turned the 4CC into the Four-Color Theorem.

FOUR COLORS SUFFICE!

Could an unavoidable set of reducible configurations be found? That was the question that haunted Wolfgang Haken, who began his career as a microwave engineer in Munich working for the firm Siemens and Halske. Following his discovery of an effective procedure for deciding whether a knot is knotted or unknotted, an old problem that had stood open for several years, in 1962 Haken was invited to the University of Illinois as a visiting professor. In 1965 he became a permanent member of the university faculty, setting the stage for his work on the 4CP, which began in earnest around 1970, when he noticed certain ways to improve discharging procedures. These developments led him to believe that just maybe it would be possible to actually prove the 4CC.

But many difficulties stood in the path to a proof. The first was that folk wisdom in the Four-Color community had it that very large configurations having rings of neighbors of up to 18 nodes would necessarily be included in any unavoidable set of reducible configurations. This meant that the problem might be beyond the bounds of then-current computing capabilities, since while it was relatively straightforward to check reducibility for small ring size up to around 11 nodes, computing time increased by a factor of four for every additional node. So a ring size of 18 would step up the computer time by a factor of about 16,000 from that needed for an 11-node ring. Computer storage requirements were another obstacle. Dürre's program showed that testing an average 18-node ring required more than 100 hours of computer time and more storage than was available on any existing computer.

A second major problem was that no one really knew how many reducible configurations would be needed to form an unavoidable set. It seemed likely that this number would be in the thousands. But no upper bounds were known. So, for instance, if it took 100 hours to show that an 18-ring configuration is reducible, and if there are 1,000 such rings in the unavoidable set, it would take 100,000 hours—11 years!—to prove them all reducible.

Haken and Appel state that when they began their work together in 1972, they were extremely pessimistic that any noncomputer-based proof of the 4CC would be found. The problem had been worked on by many famous mathematicians for over a hundred years, many approaches had been tried, but none had come close to providing anything looking like a proof. Therefore, their first step in attacking the problem was to determine if there was any hope of finding such a set of configurations having ring size small enough that the computer time needed to prove reducibility of all configurations would be within reason. Here Heesch's work reentered the picture.

During his testing for reducibility, Heesch found three major obstacles to reduction. No configuration containing one of these obstacles had ever been proved to be reducible. Furthermore, it's easy to determine whether or not a configuration contains one of these reduction obstacles, and configurations without such an obstacle have an excellent chance of being reducible. So if there existed an unavoidable set of configurations free of these obstacles, it was likely there would be an unavoidable set of roughly the same size that contained only reducible configurations. Thus, Haken and Appel first studied various types of discharging procedures in order to determine the kinds of sets of configurations that might arise that would be free of the obstacles noted by Heesch.

In autumn 1972 they wrote a computer program to carry out the type of discharging procedure that seemed most reasonable to them. They were not ready at this time to actually prove the 4CT, as the output of their program was a list of configurations that resulted from the most important situations, not an unavoidable set.

The first runs of this program gave much valuable information. First of all, it appeared that good configurations of reasonable size (ring size less than 17) would be found close to most nodes that would ultimately have positive charge. Secondly, the same configurations occurred often enough for it to appear likely that the list of configurations needing testing would be of manageable size. Other favorable features also emerged from this preliminary study, encouraging Haken and Appel to move on to the next stage of their experiment.

A man-machine dialogue then ensued for half a year or more, during which several versions of the computer program were used to deal with problems pointed out by earlier versions. Finally, they were ready to try to prove that the procedure the team was using would produce a finite unavoidable set of configurations in a reasonable amount of time. To the great surprise of Haken and Appel, it was extremely difficult to do, and it took more than a year for this proof. But by the fall of 1974, they had a rather lengthy proof that a finite unavoidable set of configurations does exist, and they had a procedure for constructing such a set, along with definite bounds for how large the set of configurations could be. But they still didn't know how complicated it would be, computationally speaking, to carry out their procedure. It may be that the computations involved might dwarf the capacity of even the supercomputer then available at the university. But as it turned out, this was not the case, at least on the test problem of triangulations having no pairs of adjacent nodes of degree-5. And by then the program was displaying surprising characteristics of its own. Previously, Haken and Appel were always able to predict the course of the computer analyses when they hand-checked the output. But no more. Now the program was working out compound strategies based on all the tricks they had taught it, and the new approaches were often much more clever than those the program's designers would have employed themselves. So on they went in their quest to settle the 4CC.

By the summer of 1975 Haken and Appel thought there was a good chance of finding an unavoidable set of configurations that would be obstacle-free and reducible. So they began to write a program for the IBM 360 computer to test configurations for reducibility. In this effort they were joined by John Koch, a graduate student in computer science, who had chosen to write a dissertation on the reducibility of configurations of small ring size. By the fall of that year Koch had written a program to check reducibility in configurations of ring size less than 12, a program that was later modified to go up to ring sizes of 14 or less. But now a new problem arose.

The discharging procedure Haken and Appel had chosen needed major structural changes, not just incremental adjustments, in order to deal with configurations of the sizes that would be needed to produce a reducible unavoidable set. So instead of writing an entirely new program to do this, to maintain flexibility and the ability to make changes on the fly, they chose to implement the discharging procedure by hand. And in December 1975 they discovered that one of the rules defining their discharging procedure was far too rigid. Relaxing this rule then greatly improved the efficiency of their procedure.

As the new year began in 1976, the Illinois team began the construction of an unavoidable set of reducible configurations using their new discharging procedure. They started by making a preliminary run of their new method of discharging, in which they considered each possible case in which a node was forced to have positive charge. In each of these cases, they examined the neighborhood of the node looking for an obstacle-free configuration. If none was found, the neighborhood was termed *critical,* meaning the discharging procedure would have to be modified to avoid the obstacle in future runs. But even in an obstacle-free configuration, they couldn't guarantee the configuration was reducibile. So the new reduction procedures were then used to try to find some obstacle-free configuration in the neighborhood. If none was found, the neighborhood was also termed critical. Again, this method of developing an unavoidable set of reducible configurations involved extensive dialogue with the computer. And to determine which neighborhoods were critical, it was necessary to be able to check for reducibility quickly. Fortunately, this turned out to be an easy requirement to satisfy, and it was seldom necessary to wait for more than a few days to get the results.

Success came in June 1976 when Haken and Appel succeeded in finding an unavoidable set of reducible configurations. The Four-Color Conjecture was now the Four-Color Theorem. To celebrate this triumph, Figure 2.13 shows how the postage meter in the Mathematics Department at the University of Illinois modified its stamp to honor the proof. In the course of constructing their proof, Haken and Appel had used more than 1,200 hours of computer time on three different machines, as well as hand analysis of some 10,000 neighborhoods of nodes having positive charge and machine analysis of more than 2,000 configurations. The final proof involved the reduction of 1,482 configurations. Although the discharging procedure sans reduction can be hand-checked in a few weeks, it seems virtually impossible to verify the reduction computations in this manner. And, in fact, when Haken and Appel submitted their paper on the proof to the *Illinois Journal of Mathematics*, the referees did check the discharging procedure by hand—but checked the reduction computations by running them on a different computer using a different program. And therein lies perhaps the most interesting aspect of the proof of the Four-Color Conjecture.

Figure 2.13. Postage meter stamp honoring the Haken-Appel proof of the 4CC.

MIXED REVIEWS

In June 1993, when Andrew Wiles announced his solution of the Fermat Problem in Cambridge, the audience erupted with a great ovation. And rightly so. They had just witnessed an historical event. When Wolfgang Haken first announced the proof of the Four-Color Conjecture at the 1976 summer meeting of the American Mathematical Society and the Mathematical Association of America, held at the University of Toronto, the lecture hall was packed. But the audience was anything but wildly enthusiastic. About the best one could say is that Haken left the podium to polite applause. Mathematician Donald Albers was puzzled about this, as he had expected Haken's talk would receive a very warm and loud reception. So after the presentation Albers circulated among his colleagues and asked them to explain the cool reaction they had given Haken's presentation. The reason was that Haken and Appel had used a computer: "Mathematician after mathematician expressed uneasiness with a proof in which a computer played a major role," is how Albers put it.

Following this public unveiling of the Four-Color result, Haken gave many lectures around the country and the world about it. He often met with sharp hostility. One mathematician, for example, tried to prevent Haken from meeting with any of the graduate students in his department. As Haken put it,

> Since the problem had been taken care of by a totally inappropriate means, no first-rate mathematician would now work any more on it, because he would not be the first one to do it, and therefore a decent proof might be delayed indefinitely. It would certainly require a first-rate mathematician to find a good, satisfactory proof, and that was not possible. So we had done something very, very bad, and things like that should not be committed again, and he had to protect the innocent souls of his students against us.

Ian Stewart of the University of Warwick in the UK, a well-known mathematician and popularizer of mathematics, expressed his own disquiet in print, complaining that it

> ... doesn't give a satisfactory explanation *why* the theorem is true. This is partly because the proof is so long that it is hard to grasp (including the computer calculations, impossible!), but mostly because it is so apparently structureless. The answer appears as a kind of monstrous coincidence. Why is there an unavoidable set of reducible configurations? The best answer at the present time is: there just is. The proof: here it is, see for yourself. The mathematician's search for hidden structure, his pattern-binding urge, is frustrated.

Perhaps symptomatic of the unease felt by the mathematical community over the proof were the continual rumors floating about that there was a "bug" in the program. So prevalent were these rumors in the early 1980s, that the editor of the *Mathematical Intelligencer* invited Appel and Haken to set the matter straight. Suffice it to say that no such bug has been discovered. Interestingly, errors *were* found in the hand-generated part of the proof, the part involving the construction of the unavoidable set. But these errors were minor and easy to fix.

Even the possibility of a bug in the reducibility program doesn't bother Appel and Haken very much. Their confidence in the correctness of the proof is supported by the fact that reducibility can be checked by independently written programs, written in different languages and run on different machines. Moreover, they feel that the very "inelegance" of their proof makes it more robust compared to proofs employing more elegant—but fragile—chains of reasoning. They don't claim to have found the only, or the best, proof of the 4CC—just one proof among many.

Lending still more support to the correctness of the result is a second proof found a year later by Frank Allaire. He used the same approach, but a different discharging procedure, and more powerful methods of reducibility. Yet a third proof was presented in 1993 by Neil Robertson, Daniel Sanders, Paul Seymour, and Robin Thomas. Again, they used a different discharging procedure than either of the previous proofs, and succeeded in constructing a reducible unavoidable set having only 633 configurations. So there is no longer any doubt that the result is correct: Four colors do suffice!

Despite the correctness of the result, some argue that Haken and Appel's work calls into question the very notion of "proof" itself. Here are the words of F. P. Bonsall, professor of Mathematics at the University of Edinburgh:

> If the solution involves computer verification of special cases ... such a solution does not belong to mathematical science at all. ... We cannot possibly achieve what I regard as the essential element of a proof—our own personal understanding—if part of the argument is hidden away in a box ... Perhaps no great harm is done, except for the waste of resources that would be better spent on live mathematicians, so long as we do not allow ourselves to be content with such quasi-proofs. We should regard them as merely a challenge to find a proper proof ... Let us avoid wasting those funds on pseudo mathematics with computers and use them instead to support a few real live mathematicians

So what is proof? Enter the philosophers.

WHEN IS A "PROOF" NOT A PROOF?

The December 20, 1988, issue of *The New York Times,* in an article titled "Is a Math Proof a Proof If No One Can Check It?" reported the results of a computer search for an object called a finite projective plane of order 10. The search, which was initiated to settle a long-standing mathematical conjecture, consumed several thousand hours of time on a Cray supercomputer, eventually rendering the verdict that there are no such planes, thus confirming the conjecture. What C. W. H. Lam, the author of this search, did not tell the media was that the Cray is reported to have undetected errors at the rate of approximately one per thousand hours of operation. So it would certainly be reasonable to expect a few errors during the course of the computer's work on this search. As Lam remarked, "Imagine the expanded headline, 'Is a Math Proof a Proof If No One Can Check It and It Contains Several Errors?' " So when it comes to giving a "good" proof, how good is good enough?

By common consensus in the mathematical world, a good proof displays three essential characteristics: a good proof is (1) *convincing,* (2) *surveyable,* and (3) *formalizable.* The first requirement means simply that most mathematicians believe it when they see it. The philosopher Ludwig Wittgenstein held to the skeptical view that in practice this was the *only* requirement for a mathematical proof, a good indication of why he was a better philosopher than a mathematician. Most mathematicians and philosophers of mathematics demand more than mere plausibility, or even belief. A proof must be able to be understood, studied, communicated, and verified by rational analysis. In short, it must be surveyable. Finally, formalizability means we can always find a suitable formal system in which an informal proof can be embedded and fleshed out into a formal proof.

These three features of a good proof reflect three very different aspects of the practice of mathematics, each of which is a crucial component of the way mathematics is actually done. That a proof must be convincing is part of the *anthropology* of mathematics, providing the key to understanding mathematics as a human activity. We invoke the *logic* of mathematics when we demand that every informal proof be capable of being formalized within the confines of a definite formal system. Finally, the *epistemology* of mathematics comes into play with the requirement that a proof be surveyable. We can't really say that we have created a genuine piece of knowledge unless it can be examined and verified by others; there are no private truths in mathematics. It is questions about surveyability that lie at the heart of the difficulty many mathematicians have in swallowing the computer-assisted proof we've just outlined of the Four-Color Conjecture. From a philosophical point of view, this is by certainly the most interesting aspect of Haken and Appel's monumental achievement.

The heart of the philosophical debate over the computer proof of the 4CC is that there is no way even an army of mathematicians working for a time longer than the age of the universe could possibly hand-check each of the reduction calculations needed to verify that Haken and Appel's unavoidable set of configurations is indeed reducible. So the proof is not surveyable, at least in the sense used in mathematics up to that time. Strangely enough, though, the part of the proof that most bothered the mathematicians was not the reducibility calculations, but the work done by hand in implementing the discharging procedure.

It is by no means an easy matter to check that the final working list of 1,482 unavoidable configurations encompasses all the unavoidable configurations resulting from the discharging procedure employed by Haken and Appel. Nevertheless, the main focus of the arguments by philosophers like Thomas Tymoczko is that the unsurveyability of the proof makes the 4CT the first "a posteriori" mathematical truth. Now what does he mean by this?

Traditionally, mathematical truths have been considered to be a priori truths, either in the sense that they are truths that would be true in any possible universe, or in the sense that they are truths whose validity is independent of our sensory impressions. Tymoczko takes the second interpretation, and argues that our sensory impressions cannot ever verify the proof of the 4CT; hence, the truth of the conjecture must be an a posteriori truth. Mathematician E. J. Swart argues to the contrary, using the following example from an early assault on the 4CC by Oystein Ore and Joel Stemple.

Stemple and Ore proved the following theorem:

Every triangulated planar graph of less than 40 nodes is 4-colorable.

To obtain this result, Ore and Stemple had to hand-check 42 separate cases to determine their reducibility. The figures and calculations comprised 118 pages of dense script, and were thus far too long for any journal, so they were simply written out (by hand!) and deposited in the Mathematics Department at Yale University, where any interested mathematician could consult them. The point is that carrying out this testing is far too cumbersome to be done in one's head, and required a substantial amount of storage space (paper!) and processing hardware (pencils) to complete. Swart argues that in practice theorems that involve testing of cases fall into four categories:

i. Those theorems for which case testing can be done in our heads.
ii. Those theorems in which the case testing is impossible to carry out without the aid of pencil and paper.
iii. Those theorems for which the testing can be carried out with immense effort by pencil and paper.
iv. Those theorem that are entirely beyond the reach of hand calculation, and for which the case testing *must* be done by computer.

These categories are not hard and fast, and a given problem may well move from one to the other as technology and mathematical wisdom develop. At the time of their proof, Ore's theorem was in category (ii), while the 4CT was in category (iv). But no one has proved that this is a permanent categorization. Thus it seems rather far-fetched to classify theorems as a priori or a posteriori on the basis of their present categorization, since the classification may change in time. This would indeed be a very odd use of the term a priori. Now what about the objection noted by Lam, that computers make mistakes and computer programs do have bugs?

Well, it is true that computer programs do sometimes have "bugs." But so do attempts to prove theorems by ordinary pencil and paper. And flaws in programs may well go unnoticed for a long time. But so do flaws in proofs done by hand, as in Heawood's discovery of the gap in Kempe's "proof" of the 4CC nearly ten years after its publication. So flaws in computer implementation are no more—or less—serious than logical flaws of any other kind.

It's also the case that sometimes algorithms that are logically sound do not produce the results they should when implemented on a computer. This is because machines are finite, and we are sometimes forced to truncate or round off irrational numbers or numbers with an infinite binary expansion. This often happens in numerical calculations, and the field of numerical analysis is devoted to overcoming these problems.

So what about the determination of reducibility on a computer? Firstly, the algorithm itself is so simple that there can be no real question about its correctness. Any doubts that arise must be with the implementation of the algorithm in the computer, not the algorithm itself. Sad to say, the computer implementation is not so simple, and the number of logical checks needed to verify the reducibility of a 13- or 14-ring configuration is very large. Nevertheless, many different programs have now been written using different testing procedures on different hardware—and all have come up with the same end result. So there is now little, if any, doubt about the validity of the Haken-Appel resolution of the 4CC.

What the proof emphasized for the first time was that mathematics is not so much different from the natural sciences as a lot of mathematicians would like to believe. And in the 25 years since the Haken and Appel work, the entirely new field of "experimental mathematics" has arisen. "Proofs" in this area have the characteristics that they are lengthy, have not been adequately formalized, and are not surveyable. Thus, they are suggestive rather than definitive. Swart suggests calling these results *agnograms,* meaning theoremlike statements that we have verified as best we can but whose truth is not known with the level of assurance we attach to theorems and about which we thus remain, to some degree, agnostic.

When logician Reuben Hersh first heard about Haken and Appel's proof of the 4CC he was very excited and eager to know more about the details. When told that the "method" boiled down to the brute-force checking by computer of a large number of different configurations, Hersh responded dejectedly, "So it just goes to show, it wasn't a good problem after all." Somehow mathematicians seem to long for more than just results from their proofs; they want insight. And the verification of a couple of thousand special cases by a computer smelled suspiciously like the kind of ad hoc exercise that ends up leading nowhere. I disagree.

In my view, the Four-Color Conjecture was a *great* problem. It forced mathematicians and philosophers of mathematics to confront head-on the nature of their profession and what we mean by mathematical "truth." Moreover, it called into question the entire notion of "proof," leading to new areas of mathematics and new modes of proof. If that's not a good problem, I don't know what is. And it would not be at all surprising to see even further dividends come to the world of mathematics from extensions and generalizations of the 4CP. Let's close the chapter by looking at just a couple of directions current work is moving.

CRAYONS, HANDLES, AND LISTS

While the 4CC may well have been settled, hardcore graph theorists continue to find fascinating variations on the theme of map coloring—and new theorems about them.

We saw earlier that in the nineteenth century Percy John Heawood solved the map-coloring problem for all surfaces except the sphere and the plane. Heawood's formula gives the *maximum* number of colors that would be needed to color a map on any surface in terms of the number of handles glued onto the surface (its genus). But not every map drawn on a surface requires a full complement of colors. So Heawood's formula is simply a sharp upper bound; a particular map may actually require far fewer colors. For instance, one color is enough for a map that has only one country! A natural question that arises, then, is if you are given a particular map on a surface, is there a quick method to determine whether that map requires the surface's full complement of colors?

Strictly speaking, the answer is No. For example, to determine if a given map on the plane can be colored using three colors is what's called an NP-complete problem. In short, it's computationally "hard." But if you exclude the plane, then the answer is Yes. In 1963, Hungarian mathematician Tibor Gallai showed that you can quickly tell whether a map needs its full complement of colors for the surface it's drawn on by checking for the appearance or nonappearance of a finite set of "bad" subnetworks within the network representing the map.

In 1994 Danish graph theorist Carsten Thomasson extended this result, proving that each surface (other than the plane/sphere) has a finite list of critical maps that require at least six colors. So if one of these critical subnetworks appear, the map will require the surface's full set of colors; otherwise, it can be colored using five colors. For instance, on the torus there are only four such critical maps, which are associated with graphs techically called $K_6, C_5 + C_3, K_2 + H_7$ and T_{11}. The network neighborhoods of these subgraphs are shown in Figure 2.14. Any map on the torus whose network doesn't contain one of these as a subnetwork can be colored with just five crayons.

So checking whether a map on the torus with N countries contains a six-color critical subnetwork requires at worst an algorithm whose computational complexity grows as the 11th power of N; it is a polynomial-time calculation. Only subnetworks with 11 nodes or less need to be checked. In principle, a list of six-color critical networks can be compiled for every surface. Once found, the list for a given surface reduces the five-colorability question for that surface to a polynomial-time calculation (one for which the computational burden grows only as a polynomial function of the number of nodes in the network). In the course of his work on the extended map-coloring problem, Thomasson realized

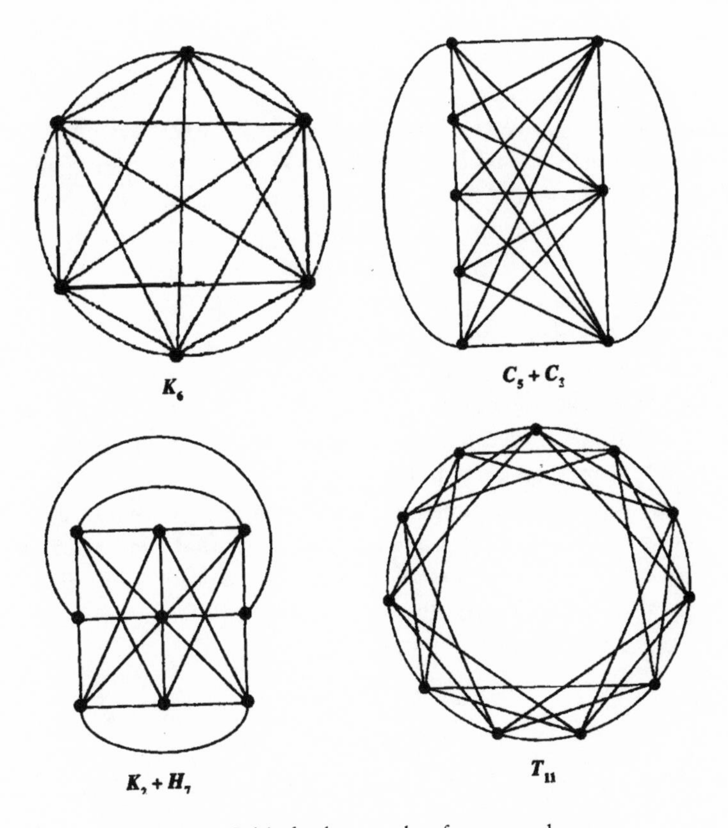

Figure 2.14. Critical subnetworks of maps on the torus.

that the methods he was using applied to another famous map-coloring puzzle, the so-called *list coloring problem.*

In 1975 Russian mathematician V. G. Vizing conjectured that if each node of a planar network were assigned a list of five colors, then the network could be colored using a color at each node taken from that node's assigned list. This problem is important in areas like scheduling of tasks, since the nodes of the network can be regarded as tasks, the lists of colors thought of as hours during the day when the tasks can be done, and the links of the network regarded as "conflicts," such as two tasks requiring the same machinery or workers. The condition that the network be a planar one corresponds to a restriction on the complexity of the conflicts.

When you first think about it, the list coloring problem sounds trivial. After all, if each node has the same list of colors, the Four-Color Theorem shows a list of four colors is sufficient to ensure that there are no planar networks that cannot be properly colored. But, of course, the catch is that the lists at each node are not the same; not every color is available at each node. So there is a problem. In fact, in 1993, Margit Voigt of the Institute for Mathematics in Ilmenau, Germany found an example of a planar network with 238 nodes and lists of four colors at each node that cannot be properly colored. This example settled an old conjecture by Paul Erdös, who believed that lists of five colors at each node would suffice, but not four.

Thomasson settled Vizing's conjecture positively, showing that indeed lists of five colors suffice to color any map, by an ingenious proof based in induction on the number of nodes. So even if the lists at each node are different, all that's needed to properly color any possible planar network is a list of five colors at each node.

On this positive note we conclude our study of maps, colors, networks, and colorings. The results discussed here show that even a seemingly trivial problem like the colors of maps can have dramatic implications for the development, even leading to the founding of entirely new fields. Who could ask for more?

SUMMARY

Four-Color Problem: Do four colors suffice to color any nondegenerate, planar map?

Answer: Yes, provided one accepts computer analysis as a "proof."

CHAPTER THREE

THE CONTINUUM HYPOTHESIS

FIRST AMONG EQUALS

In Hilbert's famous list of 23 problems (or 25, depending on how you count them) central to the development of mathematics, which he presented at the International Mathematical Congress in Paris in 1900, he reserved primacy of position on the list for a question now known as *The Continuum Hypothesis* (CH). This problem arose out of the fevered activity at the end of the nineteenth century by Georg Cantor, Gottlob Frege, Bertrand Russell, and others, who were concerned with putting mathematics on a sound, logical footing. A crucial role in these efforts revolved about how to handle the infinite. We'll look at this issue in painstaking detail later. For now, let's just consider one aspect of the overall question of infinity by considering the two most common sets of numbers: the natural numbers (the so-called "counting" numbers 1, 2, 3, ...) and the real numbers. Let's label these two sets of numbers N and R, respectively.

It's evident that there are an infinite number of natural numbers, since we can continue to enumerate the list 1, 2, 3, ... indefinitely. In fact, the three dots ... mean exactly this, "go on in the same way indefinitely." There are also an infinite number of real numbers, since each real number can be expressed as some subset of the natural numbers. For example, the real number 3/4 consists of just the two-digit sequence 75 (ignoring the location of the decimal point that is irrelevant for us in this discussion), while the real number π is the *infinite* sequence 314159265 ... Thus, the number of elements in the set N is infinite. Let's call it ∞_0 (read: "infinity zero"). Similarly, the number of real numbers is also infinite. Call this number ∞_c. The question that immediately arises is whether $\infty_0 = \infty_c$? Cantor devised an ingenious scheme to show that ∞_0 is always smaller than ∞_c. This shows that there are different "styles" of infinity. Put another way, not all infinities are created equal. We'll see how Cantor did this later. Armed with this surprising fact, Cantor asked, How many styles of infinity are there? And as a special case of this basic question, he asked if there is some style of infinity that is strictly larger than ∞_0 and strictly smaller than ∞_c? He asserted that no such type of infinity exists. This is

Cantor's Continuum Hypothesis. There is no infinity between that of the natural numbers and that of the reals.

Now why is the answer to this question so central to the development of mathematics that Hilbert would elevate it to the very first problem on his list of challenges to mathematicians for the twentieth century?

First of all, the CH is a natural question about sets. And early in the twentieth century mathematicians showed that virtually all of mathematics could be reduced to the properties of sets and tools of logical inference. Such a basic question of how many elements there are in a set thus lies right at the heart of the mathematical enterprise itself. Moreover, notions of the infinite pervade mathematics from the infinitesimals of the calculus to the continuum of time in dynamical systems and on to the infinite number of curves that can be drawn in the plane. Thus, the question of how many styles of infinity exist is about as basic a mathematical question as one can ask. But it's not just in mathematics that a question like the CH rears its head.

Much of the apparatus we use to describe the physical world is based in mathematics—Newton's equations of celestial motion, Maxwell's equations of electromagnetism, and the Schrödinger equation of quantum theory are but three examples. And in each of these cases some notion of the continuum is assumed in either time or space. But in this regard it's well to heed the words of John Archibald Wheeler, one of the great quantum theorists, who said:

> Encounter with the quantum has taught us, however, that we acquire our knowledge in bits; that the continuum is forever beyond our reach. Yet for daily work the concept of the continuum has been and will continue to be as indispensable for physics as it is for mathematics. In either field of endeavor, in any given enterprise, we can adopt the

> continuum and give up absolute logical rigor, or adopt rigor and give up the continuum, but we can't pursue both approaches at the same time in the same application.

The rest of this chapter is devoted to showing why this must necessarily be the case.

THE INFINITE

Commonplace images of the infinite seem to come in several flavors. On the one hand, there is the wall of the bottomless pit rushing past as we fall forever in a nightmare. More uplifting is the religious vision of heaven in a world without end. The notion of infinity has troubled philosophers and mathematicians since antiquity, and attempts to develop a precise theory of the infinite have been plagued by paradoxes of all sorts. For instance, think of the two concentric circles shown in Figure 3.1, where the outer circle is twice as large as the inner one. By drawing lines from the circle's common center, it's easy to pair up the points of one circle with the points of the other. Thus, the infinite number of points on the larger circle must equal the infinite number on the smaller one, even though the circumference of the larger circle is twice as long as that of the inner circle. So what gives? How can the number of points be the same, yet the lengths be different? Such is the power of the infinite!

It was not until the latter part of the nineteenth century that a decent mathematical theory of the infinite was finally put in place. This theory was invented by Georg Cantor, certainly one of the greatest geniuses in the history of mathematics. As a human interest story, it's worth spending a moment here on his tortured life. We'll come back for a more detailed look later.

Cantor was born into a highly religious family in Russia in 1845. His father converted from Judaism to Protestantism, whereas his mother was Roman Catholic, a religious mix that led to Cantor having a lifelong interest in theological issues. Some of these issues, especially those relating to the infinite, later figured heavily in Cantor's mathematics. His family was also highly focused on the arts as well as religion, as there were many painters, pianists, and violinists on his mother's side.

So this highly sensitive young man entered the University of Berlin, where his father tried to force him to be practical and study engineering. Eventually, though, his father relented and allowed Cantor to study mathematics, and he obtained his doctorate in 1867. During his studies,

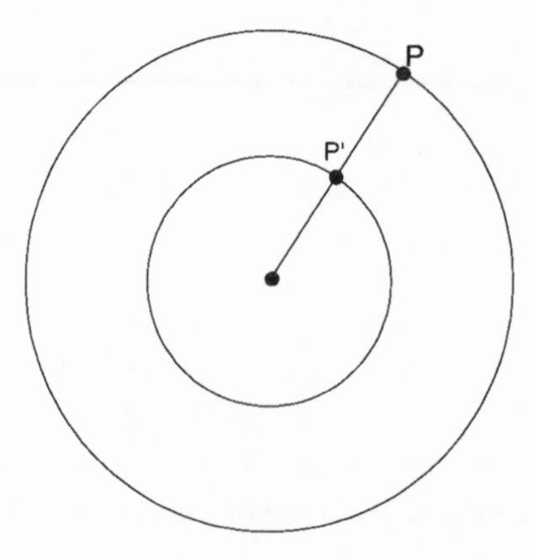

Figure 3.1. Pairing up the points of two concentric circles.

he assimilated the rigorous approach to the calculus as taught by Karl Weierstrass, which led Cantor to deeper considerations of the intrinsic differences among various sets of numbers.

Cantor's personal life was unhappy. He was unable to obtain a position at a major research university, and had to earn his living teaching at second-rate colleges. The most tragic fact of all was the violent personal and professional animosity of his teacher, the great mathematician Leopold Kronecker. Cantor lacked the stamina to be a revolutionary, taking Kronecker's criticisms so much to heart that he went out of his mind and spent the last days of his life in a mental institution.

Problems with the infinite came long before Cantor's day. More than two millennia ago Zeno of Elea, a town in western Italy, formulated a number of paradoxes involving the infinite that are taught in classes in logic to this day. Probably the most famous is the Achilles Paradox, involving the race between Achilles and the Tortise. Zeno argued that if Achilles can run 1,000 yards per minute, while the Tortise can run 100 yards per minute, Achilles will never catch the Tortise if the Tortise is given a 1,000 yard headstart. Zeno's argument is that while Achilles has advanced 1,000 yards, the Tortise has moved forward 100 yards. By the time Achilles has covered these 100 yards, the Tortise has moved forward 10 yards, and so on. Thus, the Tortise *always* remains just a bit ahead of Achilles.

The trouble with Zeno's argument is that if Achilles has to occupy an infinite number of positions in order to overtake the Tortise, how can he do this in a finite time? The solution to the "paradox" is to clarify our notions about infinity. In particular, this paradox would dissolve if we could permit an infinite number of points in a finite space and an infinite number of instants in a finite time.

More basic for mathematics is the following property of the rational and the irrational numbers. Between any two rationals we can find infinitely many irrationals and, conversely, between any two irrationals we can find infinitely many rationals. This seems to say that the real numbers must be evenly divided between these two families. As the nineteenth century drew to a close, mathematicians discovered that this was not the case at all. There must be many more irrationals than rationals. But how could one develop a way to precisely characterize this fact in a rigorous way? Enter Georg Cantor and his theory of infinite sets.

SETTING THINGS UP

The starting point of Cantor's story is with the primitive notion of a *set.* This is simply any collection of objects that share a common defining property. For example, the collection of cars currently parked on my street, all the dogs in Santa Fe, and the people I spoke with yesterday each constitute a set. If $x_1, x_2, \ldots, x_n$ are the elements of a set X, we denote this set as

$$X = \{x_1, x_2, \ldots, x_n\}.$$

So, for example, the set of people I spoke with yesterday would be

$$P = \{\text{Stu, David, Loana, Joleen, Pete}\}.$$

At the same time that Cantor was formulating his theory of infinite sets, Gottlob Frege was creating a logical structure now known as *predicate logic* to provide a universal language for expressing virtually any mathematical concept. What was important about both these developments was not that mathematicians would use them in their daily work, but rather that they demonstrated that all the branches of mathematics—analysis, geometry, algebra—are part of a single unified whole, and that this high-level language of logic enabled a proper analysis to be made of the deductive methods used by mathematicians in creating proofs of various propositions about the relations between mathematical objects. So set theory offers a suitable foundational framework upon which all mathematical objects and structures can be constructed. And predicate logic provides a universal language for defining and discussing these objects and structures. What could be better?

The fly in the ointment of set theory and predicate logic came in a letter dated June 16, 1902, from Bertrand Russell to Frege, just as the second volume of Frege's work *Foundations of Arithmetic* was going to press. Following an opening paragraph in which Russell heaped praise upon Frege's work, the letter continued: "There is just one point where I have encountered a difficulty." Russell then went on to give an explanation of a phenomenon he had discovered a year earlier—and which destroyed Frege's work entirely!

Russell's Paradox, as it has since been termed, involves the notion of sets being members of themselves. For example, the set of all dogs is not a member of itself, since it is a set, not a dog. On the other hand, the set of all nondogs is a member of itself since this is a set, not a dog. If we call the set of all sets that are not members of themselves S, then if S is a member of itself, by definition it must not be a member of itself. Similarly, if S is not a member of itself, then again by definition it must be a member of itself. A contradiction!

What made this paradox so devastating is its basic simplicity. It uses only the most fundamental concepts of set theory and logic, and so threatens the entire edifice that Cantor and Frege had created to encompass all of mathematics; hence, it threatened the logical structure of mathematics itself.

The way around Russell's Paradox was provided by Ernst Zermelo. In a paper published in 1908, he proposed a system of axioms for set theory. These axioms involve "self-evident" truths about sets, such as that the set of all subsets of a set is also a set. The self-evidence of these axioms is a little less self-evident than one might wish, and in fact it is the way we question the axioms that leads to set-theoretic paradoxes like Russell's Paradox above. As later modified and extended by Abraham Fraenkel, Zermelo's axioms came to be accepted as a "correct" axiomatic foundation for the theory of abstract sets. Within this so-called *ZF framework,* paradoxes of the type discovered by Russell do not arise, yet the framework is broad enough to encompass virtually all of the mathematics done today. Of course, in view of Gödel's results outlined in the opening chapter, there is no way to *prove* that the ZF axioms of set theory are consistent. But most mathematicians believe that they will not lead to any contradictions, an act of faith that gets stronger as each new theorem is proved within this framework without leading to logical difficulties.

The question of completeness of the ZF framework is also of crucial importance. Again, Gödel's theorem shows that there are statements about sets that cannot be settled using the ZF axiomatic setup. Until

1963 most working mathematicians felt that while this possibility exists, it was rather remote from day-to-day mathematical practice and so they more or less ignored it. But in 1963 Stanford mathematician Paul Cohen discovered problems at the foundation of set theory that simply could no longer be ignored.

Initially, Cohen's work was concerned with a problem involving how to "measure" infinity, a question that formed the original basis of Cantor's work on set theory. So let's finally move off into the weird and wonderful world of the infinite and see what's going on.

FLYING OFF TO INFINITY

The essential ingredient in Cantor's treatment of the infinite was his recognition of the importance of developing a means for comparing the sizes of two sets of elements. On the surface, comparing sets for size is a trivial consequence of counting. Suppose someone asks if you have the same number of fingers on your left hand as your right. You could just count the fingers on each hand separately, and seeing that there are five on each, answer that indeed each hand has the same number of fingers. But Cantor saw things differently.

Suppose our culture had such primitive mathematical development that people can count only up to "three." In this case we could not compare left and right hands by counting them, since our number system would not go far enough. But we could still answer the question, not by attempting to count the fingers, but just by placing our hands together, left thumb against right thumb, left index finger against right index finger and so forth. This would give a perfect one-to-one correspondence, and again we would reply that we have as many fingers on the left hand as on the right.

This example shows that we need not be able to count objects in sets in order to determine whether or not the sets have the same number of elements. *Au contraire.* The idea of being of equal size, seen in the light of one-to-one correspondences, takes on the look of something even more basic than counting. And the same idea can be employed even if the sets involved have an infinite number of elements.

Consider the sets

$$A = \{1, 2, 3\}, \qquad B = \{\sqrt{2}, \pi, 17\}.$$

Each of these sets has a finite number of elements—3. So the sets are finite.

But what about a set like the set of all natural numbers, 1, 2, 3, ... ? The property of being a natural number certainly distinguishes some numbers from others. So we can group these numbers into a collection and call it N, the set of natural numbers. Moreover, as we can continue to enumerate these elements indefinitely, this set is certainly not finite. So here is an example of an infinite set. So are the set of all integers $Z = \{0, \pm 1, \pm 2, \dots\}$ and the set of rational numbers Q, whose elements are all the ratios of integers.

It's easy to see that sets A and B have the same size, since it's possible to match their elements in such a way that each member of A is matched by one and only one element of B. We call the size of a set X the *cardinality* of X, hereafter abbreviated card X. The sets A and B both have cardinality 3. To see that they are the same size, however, we don't really have to count their elements. All that's necessary is to match them in a one-to-one fashion. For instance, we could use the matchup $1 \leftrightarrow \pi$, $2 \leftrightarrow \sqrt{2}$, $3 \leftrightarrow 17$ to do this.

An important and easy way of forming a new set from a given set X is to consider all possible subsets of X. This is called the *power set* of X and is denoted $\mathcal{P}(X)$. So, for instance, if X is the finite set consisting of the numbers 1, 2, and 3, then

$$\mathcal{P}(X) = \Big\{\{1\}, \{2\}, \{3\}, \{1,2\}, \{1,3\}, \{2,3\}, \{1,2,3\}, \{\varnothing\}\Big\}$$

(here $\{\varnothing\}$ is the empty set consisting of no elements). We see that the number of elements of $\mathcal{P}(X)$ is 8. It is a general result that if the set X has n elements, then the number of elements in $\mathcal{P}(X)$ is 2^n.

How do we measure the size of an infinite set? What is the cardinality of an infinite set like N? Is the set of natural numbers N larger than the set of real numbers c? After all, infinity is infinity, isn't it? Or is it? Somehow, the set of real numbers looks larger than the set of natural numbers, since the natural numbers are a proper subset of the reals. Similarly, the set of even numbers 2, 4, 6, ... seems smaller than the set N, even though they both contain an infinite number of elements. These simple observations led Cantor to the conclusion that we really don't want to count the absolute size of a set, but rather we want to *compare* the relative sizes. Basically, we want to be able to say that the set N is smaller than c, which in essence means that there are different sizes, or "styles," of infinity. How to do this?

Once we grasp the idea that "same size" means a one-to-one correspondence, we can develop a system of numbers that can be employed to measure the size of any set, finite or infinite. Thus, using this set of numbers, we will find that the two infinite sets

$$Z = \{1, 2, 3, \ldots\}, \qquad \text{and} \qquad E = \{2, 4, 6, \ldots\},$$

have the same "number" of elements. This number, incidentally, is usually labeled $\aleph_0$, the first letter of the Hebrew alphabet. Sets with this number of elements are often called "countable," since they can be put into one-to-one correspondence with the set of "counting numbers," N.

Cantor gave many interesting and important examples of sets that at first sight appeared to be uncountable but really are countable. For example, consider the set of positive rational numbers p/q, where p and q are positive integers. Here there are infinitely many numerators and a similar infinite list of denominators. How can we put this set into a one-to-one correspondence with the positive integers? One way to do this is indicated diagramatically in Figure 3.2.

This and other examples led Cantor to conjecture that every infinite set is countable. Indeed, Cantor spent many years trying to prove this conjecture. Then he had an *Aha!* experience. He recognized that the conjecture was false! Specifically, Cantor developed a simple, yet ingenious, scheme to show that the power set of any set is always larger than the set itself—including infinite sets. He then applied this result to the case when the set in question is the integers $Z = \{0, \pm 1, \pm 2, \ldots\}$. The power set of the integers is the real numbers, since any real number has an infinite decimal expansion consisting of a set of integers, which by definition must belong to $\mathcal{P}(Z)$.

We have just seen that the set of natural numbers N and the set of even numbers E have the same size, $\aleph_0$. Many of the sets normally used in mathematics also have this same cardinality. For instance, the set of integers, the set of rational numbers Q, the set of prime numbers, and the set of powers of two all have "size" $\aleph_0$. Cantor proved, though, that the set of real numbers is definitely bigger than this. In an 1874 paper with the unpromising title *Über eine Eigenschaft des Inbegriffes aller reellen algebraischen Zahlen* ("On a Property of the Totality of All Real Algebraic Numbers"), Cantor announced an ingenious way of establishing this fact. Since this proof is central to the entire development of Cantor's theory of the infinite, we'll go through it now in some detail.

Uncountability of the Continuum. To begin, let's agree that for this discussion a "continuum" means an interval of real numbers. So, for instance,

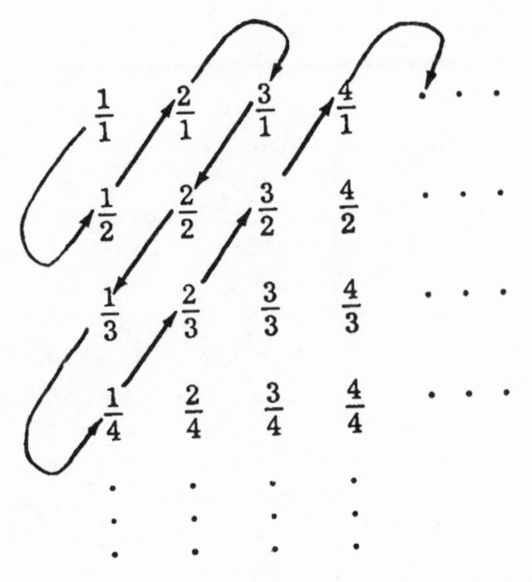

Figure 3.2. A way to "count" the positive rational numbers.

the real numbers between a and b will be denoted by the open interval (a, b). We will show how Cantor's argument establishes the uncountability of the unit interval, $(0, 1)$. Note that real numbers in this interval can all be expressed as infinite decimals. For example,

$$\frac{1}{4} = .250000\ldots, \qquad \frac{2}{3} = .666666\ldots, \qquad \frac{\pi}{4} = .7853916\ldots$$

For technical reasons, we need to avoid two different decimal representations of the same real number. For instance, $\frac{1}{2} = .50000\ldots.$ But the same number can also be written as $.49999999\ldots.$ In such cases, we will always use the expansion ending in all 0s rather than a string of nines. With this convention, all numbers in $(0, 1)$ have a unique decimal representation. Now let's turn to Cantor's proof of the following basic theorem:

Cantor's Uncountability Theorem. The interval of real numbers $(0, 1)$ is not countable.

To prove this dramatic result, Cantor assumed that the natural numbers N and the interval $(0, 1)$ can be put into a one-to-one correspondence with each other. This means that we can pair each real number in the interval with one and only one positive integer. Since each real number can be expressed as an infinite sequence of the ten natural numbers 0, 1, 2, ... , we can express a real number r between 0 and 1 as $r = .z_{r_1} z_{r_2} z_{r_3} z_{r_4} \ldots$, where each z_{r_i} is one of the ten integers between 0 and 9. By assumption, we can put any such number into one-to-one correspondence with a natural number. Suppose the following table is such a correspondence

$$r_1 = .\underline{z_{11}} z_{12} z_{13} z_{14} \ldots$$
$$r_2 = .z_{21} \underline{z_{22}} z_{23} z_{2}4 \ldots$$
$$r_3 = .z_{31} z_{32} \underline{z_{33}} z_{34} \ldots$$
$$\vdots$$

Now consider the number represented by taking the diagonal elements in this array, which are underlined, changing each diagonal element to be a digit different from the digit z_{ii} that is there initially (with the proviso that the new digit is not either 0 or 9). This procedures yields a new real number $s = .\bar{z}_{11}\bar{z}_{22}\bar{z}_{33}\bar{z}_{44}\ldots$. This number cannot be either .0000... or .9999... = 1 because of our prohibition on choosing 0s or 9s. Moreover, the number s is nowhere on the original list $r_1, r_2, r_3, \ldots$ either, since it differs from r_1 in the first position, from r_2 in the second position, and so forth. Thus, we have a contradiction, because by assumption the set $r_1, r_2, r_3, \ldots$ consisted of *all* the reals in the interval $(0, 1)$. Therefore, the unit interval cannot be placed into a one-to-one correspondence with the natural numbers, and hence has a cardinality greater than $\aleph_0$. This is Cantor's famous diagonal argument, which has been used repeatedly in mathematics ever since it was first introduced by Cantor to prove this fundamental result.

Sometimes people who hear this argument for the first time agree that Cantor found a number s that does not appear on the original list. But they then suggest the following way out: Why not simply place s opposite r_1, and then move each number on the list down one position? Cantor's contradiction seems to have been eliminated, since s now appears on the list.

Sad to say for the skeptic, Cantor could simply repeat the diagonal process on this new list, thereby obtaining a real number s' that appears nowhere on the list. The sum total of this line of reasoning is that a one-to-one matching between N and the unit interval is impossible.

It's also worth noting that there is nothing special about the unit interval in this argument. We could have used any interval of real numbers,

(a, b), by introducing the function $y = a + (b - a)x$, which is a one-to-one correspondence between the points of the interval (a, b) and the unit interval. This matching guarantees that the intervals $(0, 1)$ and (a, b) have the same cardinality, c. At first glance, it seems odd that *any* two intervals have the same number of points. But that's just the way of things with infinite sets—strange, counterintuitive things happen.

From here it's just a small step to show that the set of *all* real numbers has the same size as the unit interval, $(0, 1)$. The function $y = (2x - 1)/(x - x^2)$ is a one-to-one matching of the unit interval to the entire set of real numbers, thereby establishing this fact.

This immediately raises the question: What *is* the size c of the set of real numbers? In particular, is it $\aleph_1$, the next infinite number? Or is it something else? Cantor was unable to answer this question, as were other eminent mathematicians. Ultimately, this came to be known as Cantor's *continuum problem,* so called because it asks for the size of the "continuum" of real numbers. Another way of stating the Continuum Hypothesis, then, is to assert that $c = \aleph_1$. That is, there is no style of infinity bigger than countable and less than the continuum. The Continuum Hypothesis asserts that there is no set whose cardinality is larger than that of the natural numbers N and smaller than the power set of N, the real numbers. The so-called *Generalized Continuum Hypothesis* claims that for *every* set S, there never can be a set of higher cardinality than S but lower cardinality than $\mathcal{P}(S)$.

Here it's worth pausing for a moment to observe that the Continuum Problem has a tantalizing connection with the theory of information, which enables us to see it in a somewhat more intuitive light.

Information and the Continuum Hypothesis.

Suppose we call a sequence of 0s and 1s of length $\aleph_0$ an "infinite bit," or an ω-bit for short. It is not known whether the physical features of our world can be coded in ω bits of information. Certainly some of the most important features of the world of mathematics do code in this way. For example, Gödel's Incompleteness Theorem shows that the set of all true sentences about numbers cannot be represented in any finite way; it needs, in fact, ω^ω bits to code all such statements.

Mathematician and author, Rudy Rucker, speculates that if we view the world purely as information, it seems likely that each feature of the world does code as an ω bit of information. By this he means that given any object, we can ask an endless set of Yes-No questions about it. He further argues that while his own set of such questions is only one ω bit, there might be many endless such lines of enquiry existing independently of his own. He then concludes that the world, in each of its aspects, is not a game of 20 questions, but a game of an infinity of questions.

Cantor's argument comes into play when we speculate that there might be some one master ω bit that codes up all the world's information, if only it could be decoded in the right way. But this is exactly what Cantor's diagonal argument forbids! No decoding process, however subtle, can turn one infinite bit into *all* infinite bits.

So in terms of information theory, the Continuum Hypothesis can be restated like this: Is there any body of information that is intermediate in size between being (a) so small as to be codable by a single ω bit, and (b) so large as to code up the information in every ω bit?

Rucker speculates that a combined theory of physics and information might have something to say about this question. For example, suppose we were able to develop a theory under which the fundamental units of reality—the "particles"—are ω bits. Within the framework of such a theory one might discover a class A of phenomena such that A has too much information to be coded by any ω bit, but has essentially less information than the entire universe of ω bits. A might be, for instance, associated with a certain class of elementary particles, or with a certain class of idealized observers. Under these circumstances we would have a reason for saying that the answer to the Continuum Hypothesis is Yes. In other words, there are sizes of infinity between ω ($\aleph_0$) and the continuum, and thus c is not equal to $\aleph_1$.

From the time of the Greeks up until Cantor's period, thinkers of all persuasions had recognized only the "potential infinite." Thus, they would certainly recognize that the set of natural numbers, N, is infinite, in that we will never run out of them. But they would certainly have objected to the idea of the "completed infinite," which would imply that the process of writing down each natural number can actually be completed.

Cantor, however, was perfectly willing to regard the natural numbers as a self-contained, completed set that could be compared with other infinite sets of elements. Unlike Gauss, for example, Cantor was not ready to dismiss infinity as a mere figure of speech. To Cantor it was a totally respectable mathematical concept, every bit as tangible as finite objects like the integers between 1 and 10 or the polynomial $p(x) = x^2 + 3x + 1$.

So where did Cantor go from here?

BEYOND INFINITY

Just as there is an infinite set of finite numbers, 1, 2, 3, ... , there is also an infinite list of infinite numbers, $\aleph_0, \aleph_1, \aleph_2, \ldots$, each one "larger" than the one before. But how do we know that each one of these numbers is larger than its predecessor? That was the problem Cantor faced, the need to formalize the concept of "less than" for this new kind of number.

One might try to define the notion of less than by saying that the card A is less than the card B if there is a one-to-one correspondence between all the elements of A and *some* of the elements of B. In other words, all of A can be matched with a proper subset of B.

Sad to say, while this definition works fine for finite sets it is completely unsatisfactory when we move to infinite sets. For instance, consider the set of natural numbers N and the set of rational numbers Q. It's easy to write down a one-to-one correspondence between all of N and a small subset of Q. For instance, match the natural number n with the rational number $1/n$. Yet we certainly do not want to use this matching to conclude that the card N is less than that of Q. In fact, we have already seen in Figure 3.2 that there is a perfect one-to-one correspondence between N and Q. So both sets have the very same cardinality.

Cantor found his way out of this difficulty by introducing the notion of "less than or equal to" rather than "less than." He said that the card A is less than or equal to the card B if there exists a one-to-one correspondence from all the elements of A to a subset of the elements of B. (Note: The "subset" of B might be all of B.) With this definition in hand, Cantor could define what he meant by a strict inequality between the cardinalities of two sets. He said that card A is less than card B if card A is less than or equal to card B and there is no one-to-one correspondence between A and B. Note here that to show card $A <$ card B, we must first find a one-to-one correspondence between all of A and a part of B (establishing that card $A \leq$ card B), and then show that there can be no one-to-one correspondence between all of A and all of B.

To see this definition in action, let's use it to show that card $N = \aleph_0 < c$. For a one-to-one matchup between N and the reals, consider associating the natural number n with the real number $1/n\pi$. Hence, card $N \leq$ card R (more properly, the interval (0, 1)). But Cantor's diagonal proof showed that no one-to-one correspondence exists between these two sets. Thus card $N \neq$ card $(0, 1)$. Taken together, these facts lead to the conclusion that card $N = \aleph_0 <$ card $(0, 1) = c$.

Now having the ability to compare cardinalities, Cantor introduced a very important proposition: If card $A \leq$ card B and if card $B \leq$ card A, then card A = card B. In other words, Cantor was trying to say that if there is a one-to-one matching between all of *A* and a part of *B*, and there is an analogous matching between all of *B* and a part of *A*, then there is a one-to-one correspondence between all of *A* and all of *B*.

Cantor was completely convinced of the truth of this assertion—but could never prove it. Fortunately, Cantor's faith was justified when in 1896 the theorem was proved by Ernst Schröder and, independently, in 1898 by Felix Bernstein. Today, the result is known as the Schröder-Bernstein Theorem, which is of great power in proving interesting facts about infinite sets. Here we follow William Dunham's treatment of how the Schröder-Bernstein Theorem can be used to determine the cardinality of $\mathcal{I}$, the set of irrational numbers.

To "measure" the irrationals, Cantor needed one more basic result. Suppose *A* and *B* are both countably infinite sets, and that *C* is the set of all elements that are in either *A* or *B* (or both). Then Cantor showed that *C* is also countably infinite. For example, take *A* to be the set of even integers and *B* to be the odd integers, then the union of these two sets, *C*, is the totality of integers, the set *Z*, which we already know has cardinality $\aleph_0$.

Now let *A* be the set of rational numbers, which we already know have cardinality $\aleph_0$. Let's *assume* that the irrational numbers are also countably infinite, just like the rationals. Then the union of these two sets should also be countably infinite. But the union of the rationals and the irrationals is simply the set of all real numbers, which we know is not countable. Thus, our original assumption about the irrationals is wrong; they are not countable and, thus, cannot be put into a one-to-one correspondence with the natural numbers, *N*.

This result means that there are far more irrationals than rationals. The fact that the real numbers vastly outnumber the rationals can now be seen as a consequence of the overwhelming abundance of the irrational numbers. Therefore, even though the rationals are "densely distributed" among the real numbers, in the sense that given any distance whatsoever away from a particular real number, there is a rational number within that distance, the rationals are still just a drop in the bucket in the set of all real numbers. Almost all reals are irrational.

Cantor's magnum opus of 1874 contained one more bombshell. Having shown the uncountability of intervals, the paper now addressed a difficult question that had long exasperated mathematicians—the existence of transcendental numbers.

There are many ways to split the set of real numbers. We just did it by the rational/irrational split. But they can also be split along a different line of cleavage, one focusing on the solution to algebraic equations. Specifically, we have algebraic numbers, those reals that are the root of a polynomial equation

$$x^n + a_1x^{n-1} + a_2x^{n-2} + \cdots + a_n = 0,$$

where the coefficients a_i are integers. For instance, $\sqrt{2}$ is an algebraic number because it is the solution of the algebraic equation $x^2 - 2 = 0$. But π does not seem to be the root of any algebraic equation, and thus is what's called a *transcendental* number. The real numbers can be split along this divide into those that are roots of algebraic equations, the algebraic numbers, and those numbers that are not, the transcendentals.

The algebraic numbers seem to constitute a very large set. All the rationals are in it, as well as a lot of irrationals like $\sqrt{2}$. The transcendental numbers, by way of contrast, are hard to find. In fact, the first concrete example of one was constructed by Joseph Liouville in 1844. Thirty years later when Cantor looked at the subject, Lindemann's proof that π was indeed transcendental was still nearly ten years into the future. Many at the time thought that perhaps the transcendentals were oddballs, the exception among the real numbers.

Cantor's first step was to prove that the set of algebraic numbers was countable. With this result in hand, he considered an arbitrary interval (a, b) of real numbers. Knowing that the algebraic numbers within this interval formed a countable set, if the transcendental numbers in the interval were likewise countable, then by the result used earlier on the union of countable sets being countable, he could conclude that the entire interval was countable. But Cantor had already shown that intervals of real numbers are not countable. Hence, the transcendental numbers must constitute an uncountable subset of any interval.

Put another way, Cantor knew there were far more real numbers than could be accounted for by the relatively small set of algebraic numbers. Where did all these other real numbers come from? The only thing they could be is the mysterious transcendental numbers. So like the postulated "dark matter" of the universe that physicists believe constitutes the mass needed to keep the universe from flying off to infinity, the transcendental numbers keep the real numbers "together," so to speak—even though Cantor could not find a single, concrete example of one of them.

To many mathematical conservatives of the time, Cantor's results had the air of heresy of the first magnitude. Asserting the abundance of the transcendentals without exhibiting a single example was the height of sheer folly. Cantor heard these criticisms of his work—and it almost destroyed him! As a case study in the sociopsychology of the mathematical world, it's worth digressing for a few pages to describe in more detail Cantor's tortured life following his great work.

THE NOT-SO-LOYAL OPPOSITION

Perhaps not surprisingly, given the deeply religious nature of his family upbringing, Cantor came to find a religious significance in his theory of transfinite numbers. In fact, he regarded himself "not only as God's messenger, accurately recording, reporting, and transmitting the newly revealed theory of the transfinite numbers but as God's ambassador as well." He went on to claim that he recognized the truth of the transfinities with God's help.

In today's jargon, Cantor would have been regarded as a "religious nut case" and a "weirdo," whose attitude certainly did little to endear him to his critics. Cantor's dogmatic religious views made him an easy target for ad hominem arguments against his work. And he definitely didn't help his image when he added a fervent interest in proving that Francis Bacon wrote Shakespeare's plays to his fanatical obsession with religious questions. Yes, Cantor was definitely an eccentric. Yet there remained his mathematics.

Conservative members of the German mathematical community and elsewhere argued strongly against his work, and bad blood developed between Cantor and a number of influential mathematicians. In fairness, it must be said that not all of these objections were baseless. Cantor's mathematics raised truly puzzling questions that troubled even mathematicians of the nonreactionary persuasion.

One of Cantor's fiercest critics was Leopold Kronecker (1823–1891), a professor at the prestigious University of Berlin, where Cantor himself had studied under the famous mathematician Karl Weierstrass. Cantor's professional life, on the other hand, was spent at the University of Halle, an institution of vastly less prestige than Berlin. Cantor felt keenly the slight of being buried in such an academic backwater, and often attributed this sad fate to Kronecker's persecutions and maneuverings against him. Strongly-worded attacks flew back and forth between Cantor and his oppressors, Cantor finally coming to show distinct paranoid tendencies.

Grappling with the infinite is a tricky enough business even today, when we have had over a century to get used to it. Cantor did it alone in a hostile environment, to say the least. So it is perhaps no surprise that he suffered several bouts of mental illness and institutionalization. Cantor's first breakdown was in 1884, when he was working night and day on the Continuum Hypothesis. It's popularly believed that the stressful nature of this mathematical work, coupled with the persecution by Kronecker, were responsible for this collapse. A more thorough examination of the medical data, however, suggests that the real reason is more likely that Cantor was a manic-depressive, and that nervous breakdowns would have occurred in any event. Thus, his attacks of mental illness and depression may have been triggered by personal and professional difficulties, but they appear to have been of a more fundamental nature than just this.

Cantor's bouts of mental instability continued and became more frequent. Following a brief hospitalization in 1884, he recovered but remained fearful that the disease would return. Then his son Rudolf died in 1899, which sent Cantor over the edge again and back to the neuropathic hospital in Halle in 1902, and again in 1904, 1907, and 1911. And even when he was not in the hospital, he often had long periods at home where he simply sat, immobile and silent. This troubled life finally came to an end on January 6, 1918, while Cantor was again hospitalized.

Mathematical biographer William Dunham draws a parallel between the lives and works of Cantor and his contemporary from the world of art, Vincent van Gogh. Not only were the two men physically similar—balding, narrow-faced, and a short van-Dyke-type beard—they had very similar family backgrounds. Cantor's father was highly religious, and van Gogh's was a Dutch clergyman. Both were drawn to artistic enterprises, enjoyed literature, and wrote poetry. Like Cantor, van Gogh had an erratic, volatile personality that eventually even alienated his friends. And, of course, both Cantor and van Gogh suffered nervous breakdowns and had to be hospitalized for mental illness.

Most importantly, though, Cantor and van Gogh were revolutionaries. Just as van Gogh managed to carry art beyond its impressionistic boundaries, Cantor's work moved mathematics in fundamentally new directions. Cantor opened up entirely new vistas of mathematics, and it is perhaps fitting to note that his work has been called the first *truly* original mathematics since the Greeks.

But this first truly new mathematics already generates a truly new puzzle: How many "styles" of infinity are there, and in particular does there exist a style in between that of the integers and the reals? This is what the Continuum Hypothesis is about and we have now seen how it arose. So let's move on to an account of how work in the second part of the twentieth century finally led to its surprising resolution.

YES, NO, MAYBE

The first major advance in resolving the Continuum Hypothesis was made in 1938 by Kurt Gödel. Working within the standard ZF axiomatic framework, he showed that both the CH and the Generalized Continuum Hypothesis (GCH) were consistent with ZF. This means that no contradiction will arise within this framework if we assume that the CH is true. Thus, CH cannot be disproved within ZF. Strangely, despite this proof, Gödel stated that he believed the CH to actually be false, and that this would be established when mathematicians found the "right" new axioms with which to extend the ZF system. We'll return to this point later. To understand Gödel's argument, we need to consider the idea of a *model* for an axiomatic framework.

The general idea of a model is to provide a concrete example of a mathematical framework that satisfies the axioms and relations of an abstract mathematical theory. Just to fix this notion, here is an example of a model due to James Brown for a particularly primitive theory T. The objects of T are of two types, called *ponks* and *lonks.* There is one relation between these objects, called *zonks.* Theory T has three axioms:

1. For any two lonks, at most one ponk zonks both.
2. For any two ponks, exactly one lonk zonks both.
3. There are at least three ponks that zonk each lonk.

Now all this sounds like the rantings of a schizophrenic—and it is. But we might want to know if they are *consistent* rantings. Perhaps we can derive some conclusion like: "There is a lonk with exactly one ponk which zonks it," thereby contradicting the third axiom. The concrete model shown in Figure 3.3 shows that T is actually a consistent theory.

This model, sometimes termed the *seven-point geometry,* consists of exactly seven points (the white circles) and exactly seven lines (considering

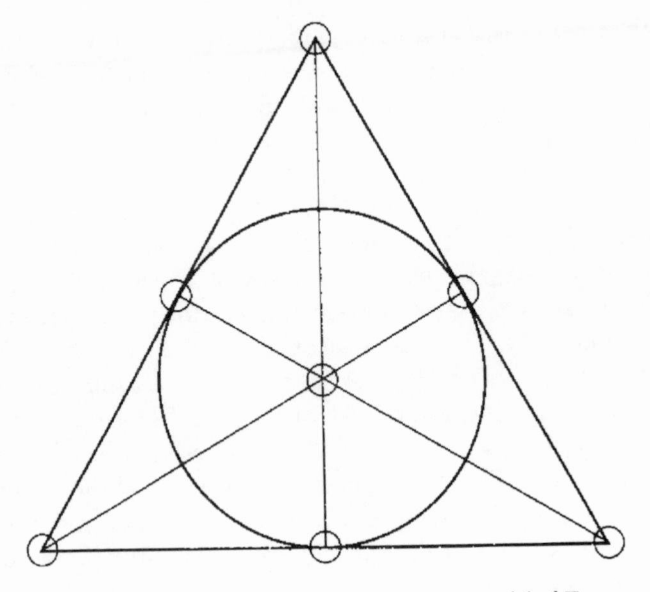

Figure 3.3. The seven-point geometry as a model of T.

the circle to be a line). We interpret the objects of the theory T as follows:

ponk = one of the seven points
lonk = one of the seven lines
zonk = the relation *lies on or connects*

With this interpretation the axioms of the theory T are all true. For instance, each pair of lines has exactly one point in common. This shows that the first axiom is true under the foregoing interpretation. Similarly, we can show the other axioms are also true. Because logical inference preserves truth, no falsehood can be derived from these axioms; hence, no contradictions can arise and thus T is a consistent theory.

In showing the consistency of T, the meaning of the terms "ponk," "lonk," and "zonk" (aside from their interpretation) plays no role whatsoever in establishing the consistency of T.

The seven-point geometry is what is termed a model for the theory T. It is a concrete mathematical structure whose elements mirror, or model, the abstract axioms and relations of the mathematical theory. With this idea in mind, we can now return to Gödel's proof of the consistency of the Continuum Hypothesis with the ZF axiomatic framework for mathematics.

Gödel proved the consistency of the Generalized Continuum Hypothesis with ZF by defining a certain type of set called a *constructible* set. Essentially, a constructible set is one that belongs to every model of ZF that is the same size as the universe of all sets. Such sets have several interesting properties, among which the most relevant for us here are these two:

1. If we interpret the word "set" to mean "constructible set," all the axioms of ZF remain true under this new interpretation. We can then say that the constructible sets form a model for ZF. Let's denote this model by the symbol $\mathcal{L}$.

   Under this new interpretation of "set," various notions of set theory might change in meaning. A notion is called *absolute* if it does not undergo such a change. As it turns out, a very large number of notions in set theory turn out to be absolute. In particular, the notion of constructibility itself is absolute. This means that constructible sets are not only constructible but are even constructible in the new sense. So the very statement "all sets are constructible," the so-called *axiom of constructibility,* is true in the model $\mathcal{L}$. Thus we have a model $\mathcal{L}$ in which both the axioms of ZF, as well as the axiom of constructibility hold. This shows that the axiom of constructibility is consistent with the other axioms of ZF, since there is a model for both (assuming, of course, that ZF itself is consistent).
2. The axiom of constructibility implies the Generalized Continuum Hypothesis. But since the axiom of constructibility is true in the model $\mathcal{L}$, so is the Continuum Hypothesis. In other words, if we reinterpret "set" to mean "constructible set," then we see that the GCH is consistent relative to ZF; it cannot be disproved using the axioms of ZF.

In 1963, Paul J. Cohen of Stanford University stunned the mathematical world by showing that the CH cannot be *proved* within ZF. More precisely, Cohen's results established the consistency of the *negation* of the CH by giving a model of ZF in which the CH is false. He managed to do this using a very powerful idea called *forcing.* Unfortunately, the method of forcing is a bit too technical to go into here. The interested reader is invited to consult the material given in the references for this chapter for the details. Cohen also gave a model for ZF in which the axiom of constructibility is false. Thus, the axiom of constructibility, like the Generalized Continuum Hypothesis, is independent of the axioms of ZF.

What all this adds up to is that the truth or falsity of the CH is simply unknown. All we can say is that the matter is undecidable within the axiomatic framework of ZF. To settle the issue, new axioms will have to be defined so as to extend ZF to a new system in which the CH can be proved or disproved. This observation leads immediately to the question of what really constitutes mathematical "truth." Let's examine this issue now, and then return to a final consideration of various ways people have proposed to extend ZF to encompass a definitive resolution of the CH.

THE TRUTH, THE WHOLE TRUTH, AND THE MATHEMATICIAN'S TRUTH

A while back, the scientific press carried a small filler piece alerting the world to the fact that researchers at the computer-manufacturing firm Amdahl had discovered that the number $391,581 \times 2^{216,193} - 1$ is prime. In addition to being the largest prime number yet found, this number is interesting from several standpoints, not the least of which is that it represents a *very* large, computable, nonrandom number. When I mentioned this achievement during the course of a lecture, one of the students in the back row looked up from his newspaper long enough to mutter, "So what? Why should anybody spend time looking for something as useless as the largest prime number? Who cares?" In response, after noting the fact that looking for such numbers is an excellent way for computer circuit designers to check the performance and accuracy of their experimental hardware, I went on to say that people look for such numbers for the same reason that other people climb mountains—because they're there. Or are they? Can we really say that the 65,087 decimal-digit "Amdahl number" is as real as, say, Mount Everest? Just what kind of a reality does a number like this actually have?

Unless you happen to be an unreconstructed solipsist, Mount Everest is definitely there; hence, certainly real. You can fall from it, die from cold and oxygen starvation on it, or climb it. There's little, if any, doubt about its reality. But numbers, or more generally, mathematical constructions and proofs, seem to enjoy a different kind of existence. You can't touch them, weigh them, or look at them. Of course, you can write numbers down in symbols and interact physically with the symbols in various ways. But in this case the medium is definitely *not* the message; the symbol is not the number being symbolized. So what is the nature of the kind of reality that mathematicians believe in? In short, how do mathematical results relate to anything that John and Jane Q. Public might think of as being "real?"

There is probably no number other than π that's more important and pops up in more places in mathematics than e, the base of natural logarithms. Its decimal expansion is $e = 2.718281828459045\ldots$, a sequence that continues forever with no apparent pattern. Now consider the following statement about the number e:

"The $10^{1,000}$th digit of e is a 4." (A)

I'd be willing to bet all of my pension (which isn't saying much) that we'll never know if this statement is true. Nevertheless, the calculation could be carried out, at least in principle, and the answer obtained. I'm equally confident that most mathematicians would say that our inability to resolve the matter is due solely to our ignorance, and that *in reality* either the $10^{1,000}$th digit of e is or is not a 4.

Next consider the following claim:

"The decimal expansion of the number e contains a run of $10^{1,000}$ 4s." (B)

In this case we again have a computational procedure to verify the statement if it is true (just keep calculating the digits of e until the required string of 4s appears), but no procedure to verify its falsity.

Finally, consider the assertion:

"The expansion of e contains an infinite number of 4s." (C)

For this statement there is no finite computational procedure available for checking whether it's true or false.

So we have three statements A, B, and C making claims about properties of the number e. As we move from statement A to statement C, we move progressively from a situation in which just about everyone would agree that the statement is definitely true or false to one in which there are pretty strongly held, mutually contradictory views about the nature of the mathematical reality of the claimed sequence of 4s. To put it simply, if it could be established that statement C is undecidable, would you still *believe* that it's either true or false? In the words of Berkeley mathematician David Gale, "At what point does God leave off and man take over?"

Beginning with the set-theoretic paradoxes of Georg Cantor in the latter part of the nineteenth century, and further stimulated by the logical paradoxes of Bertrand Russell, mathematicians and philosophers began to ponder seriously the nature of what constitutes mathematical reality. To make contact with Gale's boundary between God and man, let's take a quick tour of the major positions, using Cantor's introduction of different "styles" of infinity as a touchstone upon which to distinguish the view of mathematical reality that each competing philosophy generates.

The philosopher and mathematician Nicholas Goodman has identified the central concern of the philosophy of mathematics as being to identify that which is "practically real" in the operational experience of mathematicians. To this end, Goodman invokes what he terms the Principle of Objectivity: If a concept X plays an important role in a theory and if failure to acknowledge the role of X severely limits the theory, then X is practically real. At issue is the degree to which something that is practically real is objectively real. For example, in physics we talk about objects like quarks and strings that no one has ever observed directly. Yet their omission from the theories of elementary-particle physicists would deal these theories a severe blow. Hence, they are practically real. Moreover, a large number of physicists believe they are objectively real as well. Similarly, in geometry we deal with objects like triangles and circles. No one has ever seen a perfect triangle or the ideal circle. Nevertheless, geometry would be in bad shape if we couldn't make use of these objects. Consequently, they, too, are practically real. Are they also objectively real? It depends on whom you ask.

Historically, there are four main positions, or schools of thought, on what's objectively real in mathematics: *formalism, logicism, platonism,* and *intuitionism.* Each of these philosophical schools is based on certain foundational aspects of mathematical experience. To get a feel for the spectrum of possibilities for mathematical reality, let's take a brief look at the tenets of each group, along with its position regarding the Continuum Hypothesis.

- *Formalism:* We have already explored the formalist creed in some detail, consisting as it does of the view that mathematics is the formal manipulation of content-free symbols. Thus formalism is based on the idea of transformation rules for the shuffling-about of strings of symbols. In its current guise, formalism differs a bit from Hilbert's original conception in that Hilbert at least took the finite, combinatorial part of mathematics as being "true." But today a strict formalist would argue that the Pythagorean Theorem, for example, has no objective reality at all; it's just a string of meaningless symbols.

Continuum Hypothesis: There is no real number system outside what we create with our axioms. So, since there is no mathematical reality independent of these axioms, the Continuum Hypothesis is true or false only insofar as it can be settled within our current "best" axiomatic framework.

- *Logicism:* Logicists, like Bertrand Russell, hold to the position that mathematics consists of the kind of truths derivable from logic by an interpreted formal system of rules. They deny that these truths are "about" anything; consequently, for a logicist mathematical truths must be true solely as a result of their internal structure and of their relations to one another. Of course, in such a view there is no room for anything like intuition, since a logicist would claim that there are no structures or concepts for the mathematician to have insight into.

  Continuum Hypothesis: Basically, logicists hold to similar views as the formalists. The existence of a style of infinity between $\aleph_0$ and $\mathfrak{c}$ is purely a question of whether such a level of infinity "fits in" with our existing logical structures and methods of inference.

- *Intuitionism:* This school, sometimes called *constructivism,* is a countermovement to logicism, originally led by the Dutch topologist L. E. J. Brouwer. The basic view is that mathematics consists of intuitive constructions and of the manipulations (mentally) of symbols representing these constructions. Intuitionists accept the natural numbers as intuitively real. Every other mathematical concept or notion must be generated explicitly by a construction (read: algorithm) in order for it to be admissible. In particular, no argument based upon unrestricted use of the Law of the Excluded Middle is acceptable to an intuitionist; assertions may be true, false, or undecided. And, in general, statements are undecided unless they can be rendered true or false by an explicit construction making no appeal to *reductio ad absurdum* arguments. None of the statements A, B, or C given earlier about the properties of the decimal expansion of the number *e* would be true or false for an intuitionist until an actual calculation settled the matter. In particular, intuitionists would claim that statement C is forever undecidable, or even meaningless, since it cannot be settled by any conceivable constructive procedure. Thus, intuitionism asserts that mathematical objects exist only as a result of constructions starting with the natural numbers, and mathematical facts are true only if they follow from the results of such constructions.

> *Continuum Hypothesis:* The question is meaningless. To speak of the existence of a level of infinity intermediate between that of the integers and the real numbers is without meaning until we produce a constructive procedure to "build" an infinite set having the asserted "style" of infinity.

- *Platonism:* In Plato's philosophy the objects and notions of the phenomenological world are regarded as mere "shadows" on the wall of a cave, shadows cast by ideal, abstract objects inhabiting a universe outside ordinary space and time. But for Plato these ideal objects outside space and time were even more real than the more familiar objects of our physical and mental experience.

  Mathematical platonists like Gödel, René Thom, and Roger Penrose regard mathematical objects as also being inhabitants of this Platonic world. For such mathematicians, the content of mathematics consists of truths about these abstract structures, of the logical arguments establishing those truths, of the constructions underlying those arguments, and of the formal manipulation of symbols expressing such arguments and truths. There is nothing else. To a platonist, the Pythagorean Theorem would be a literal statement about the lengths of the sides of an idealized right triangle existing in the Platonic realm. The reality of this ideal triangle is just as real as, say, the reality of the physicist's quark or the sociobiologist's "selfish gene." And how do mathematicians make contact with this ethereal realm? According to Gödel, contact comes via the development of a sense of mathematical intuition.

> *Continuum Hypothesis:* Such a level of infinity either definitely exists or definitely does not exist in the Platonic realm. If we can't resolve the matter within our current axiomatic framework, then those axioms are incomplete as a description of the set of real numbers. In short, they aren't powerful enough to tell us the whole truth, and we should direct attention to beefing up our axioms by cultivating our intuition more intensively.

Before summarizing these various positions, it's worth once again recalling that within the framework of the commonly accepted axioms of modern mathematics, the Continuum Hypothesis is truly undecidable. As early as 1937 Gödel himself showed that the Hypothesis cannot be disproved using these axioms, and in 1963 Paul Cohen of Stanford demonstrated that it cannot be proved either. Thus the question can have a definite resolution only by extending and/or modifying the axiomatic system employed by most working mathematicians today. While many suggestions have been made as to how this might be done, no clear-cut consensus has emerged as yet showing that any of these candidates is superior to the standard system. That's the competition.

A large part of the difficulty with these philosophies is that, as a foundation for mathematics, each of them claims to be exhaustive. Moreover, each founders on a perceived need to see mathematics as being infallible, or, to put it more explicitly, to hold that mathematics generates knowledge that is certain, objective, and eternal. In view of the great difficulties each of these schools runs into when trying to shelter this collection of absolutes beneath its limited umbrella, mathematically inclined philosophers and philosophically oriented mathematicians have in recent times begun exploring the radical view that perhaps there is no fundamental difference between the practice of mathematics and that of the natural sciences. In short, mathematics is an *empirical* activity, and any viable philosophy for its foundations must respect this fact. Following up the implications of this claim leads us to a consideration of what's been termed the *quasi-empirical* approach to the foundations of mathematics.

During the Hungarian uprising of 1956, a ranking official in the Hungarian Ministry of Education fled across the border into Austria, later making his way to England. Eventually settling in Cambridge, this political refugee took up pursuit of a doctoral degree in philosophy, developing a dissertation around the theme that mathematics, like history, is fallible, and grows with criticism and correction of theories that are never totally free of the possibility of error. Unfortunately, the author of this novel thesis, Imre Lakatos, didn't live to see his work form the basis of a major upheaval in the philosophy-of-science community, particularly the mathematical branch, since he died in 1974 of a brain tumor at the relatively young age of 52. Lakatos's ideas about the practice of mathematics were published posthumously in the 1976 volume *Proofs and Refutations,* which outlined a totally new view of what constitutes a mathematical proof.

The core of Lakatos's philosophy of mathematics is that the "proof" of a conjecture means giving explanations and various sorts of elaborations making the conjecture more and more plausible and convincing. At the same time, the mathematician sharpens and refines the conjecture by subjecting it to the pressure of looking for counterexamples that would falsify it. So, in Lakatos's view, the practice of mathematics constitutes a process of conjecture, refutation, growth, and discovery, a view having much in common with the ideas of Karl Popper about the ins and outs of the scientific enterprise in general.

This "quasi-empirical" vision of the mathematical process sees mathematicians using methods analogous to those of the physical scientist, except that conclusions are verified by a proof instead of being the product of observations. For example, consider the Fermat Conjecture, which claims that there are no positive solutions to the Diophantine equation $x^n + y^n = z^n$ for any values of the integer $n > 2$. If we agree to accept the Conjecture because computers have failed to find a counterexample, then we would say that the Fermat Conjecture has been verified, or confirmed, by quasi-empirical methods.

It's crucial for Lakatos's arguments that he is able to find potential falsifiers for mathematical theories, beyond the usual logical falsifiers like inconsistency. In the natural sciences such potential falsifiers are the "hard facts" of measurement and observation. Lakatos makes the suggestion that the theorems of informal, i.e., quasi-empirical, mathematics can serve as falsifiers for more formal theories. Thus, computer investigations of the Fermat Conjecture would serve as a potential falsifier for an attempted disproof of the Conjecture by more formal procedures.

In order to accept the quasi-empirical view of mathematical truth in which a particular proposition of pure mathematics is established by appeal to empirical evidence, we must abandon or at least drastically modify certain traditional beliefs about mathematics. Some of the beliefs called into question are:

A. All mathematical theorems are known a priori.
B. Mathematics, as opposed to the physical sciences, has no empirical content.
C. Mathematics relies only on proofs, while the natural sciences make use of experiments.
D. Mathematical theorems are certain to a degree that no fact of the natural sciences can match.

Dismissal of point A is a direct rejection of the platonist position, while denial of point B strikes a severe blow against both the formalist and the logicist views of mathematical reality. Interestingly, rejection of points C and D seems to my eye to support in a vague kind of way some of the arguments of the intuitionists, but not their notion of construction as the only admissible type of mathematical proof.

A striking illustration of the shifting views of what constitutes a proof in mathematics arose in 1976 with the computer-assisted "proof" by Kenneth Appel and Wolfgang Haken of the famous Four-Color Conjecture, which we discussed in detail in the second chapter. This celebrated conjecture involved the claim that no more than four colors are needed to color any map drawn on a planar surface so that two adjacent countries having more than a single point in common (thus excluding contact like that between Arizona and Colorado) do not share the same color. The Four-Color Conjecture had resisted the efforts of mathematicians for over a hundred years, and was arguably the most famous unsolved problem in mathematics at the time of Appel and Haken's work. In Chapter Two we recounted how mathematician, Rueben Hersh, when told about this proof, felt that the proof showed that the Four-Color Problem just wasn't a very good problem, after all.

Presumably, Hersh felt this way because this method of proof didn't lead anywhere. The Four-Color Problem itself was of little intrinsic significance. After all, what real-life mapmaker really cared about whether four colors was enough for any map or not? What was thought to be important about the problem is that the proof or disproof would require the development of methods that could be used to shed light on other, far more important, questions in mathematics. But as Hersh's lament shows, the computer-aided proof by Haken and Appel was a "one-off" *tour de force,* that did not shed light on anything other than the specific problem for which it was developed.

Hersh has been one of the most outspoken advocates of an intuition-based philosophy of mathematics. In a variety of books and articles he's argued that *all* the standard philosophical viewpoints considered above make use of intuition in an essential way. For instance, platonism regards mathematical objects as already existing things, living in some ideal, timeless realm. But Hersh echoes Gödel's view when he says that we need intuition to make contact with this nonmaterial reality. Similarly, formalists, who claim that only the rules of the game matter in mathematics and that mathematical objects have no "meaning," need some kind of intuition to infuse those very same objects with semantic content. As a result of such considerations, the fundamental question of mathematical epistemology becomes: What is mathematical intuition? An answer will be forthcoming only when philosophers and mathematicians free mathematics from the shackles of papal-like infallibility.

Having seen how these various philosophies of mathematics regard the Continuum Hypothesis, let's conclude the chapter by examining some of the leading arguments put forth for why the CH is true—and why it isn't!

PROS AND CONS

In his famous 1947 paper, "What is Cantor's Continuum Hypothesis?," Gödel stated

> I believe that adding up all that has been said one has good reason to suspect that the role of the continuum problem in set theory will be to lead to the discovery of new axioms which will make it possible to disprove Cantor's conjecture.

So here is Gödel saying, in effect, that ZF doesn't represent the last word in axiomatic systems, and that there are new, currently unknown, axioms that are "reasonable" and that will enable us to disprove the CH. At one stage he even believed he had a new set of axioms and a proof that the cardinality of the continuum equaled $\aleph_2$. But this turned out to be false.

Following Gödel's work, numerous mathematicians have expressed opposing views on the matter of whether the CH is "really" true or not. Here is a short list of the pros and cons of the matter.

1. Gödel's proof of the consistency of the CH with ZF follows from the Axiom of Constructibility. Most set theorists appear to find constructibility implausible and much too restrictive. It is an example of a "minimizing" principle, which tends to cut down the number of sets admitted to one's universe. It seems that "maximizing" principles fare much better with set theorists. And such principles tend to be much more compatible with the CH being false than with it being true.
2. If the Generalized Continuum Hypothesis is true, this implies that $\aleph_0$, the cardinality of the natural numbers, has the property that it is that cardinal number before which GCH is false and after which it is true. Many set theorists believe that the the set-theoretic universe is more "uniform" than that, without this singular occurrence. Such a uniformity principle tends to imply that the CH is false.
3. Most mathematicians who think the CH is false believe that the continuum has very large cardinality, not just $\aleph_2$, as Gödel seemed to suggest. Even Paul Cohen, a diehard formalist, argues that the operation of forming the power set should yield sets much larger than those reached quickly by stepping through the ordinal numbers. He states, "This point of view regards the continuum as an incredibly rich set given to us by a bold new axiom, which can never be approached by any piecemeal process of construction."

4. In favor of the truth of the CH, we have that it is much easier to use Cohen's forcing method if the CH is true than if it's false. Moreover, the CH being true is much more likely to settle various outstanding conjectures than if it's false.
5. Most large cardinal axioms, asserting the existence of huge sets, don't seem to settle the CH one way or the other.

With these conflicting positions and intuitions about the "true" nature of the solution to the CH, one can wonder if it is even reasonable to suppose that the CH has a determinate truth value. Formalists argue that we may choose to make it come out whichever way we want, depending on the axiomatic system we work in. On the other hand, the mere fact of its independence from ZF shouldn't immediately lead us to this conclusion, as this would be assigning ZF a privileged status that it hasn't necessarily earned. After all, various axioms within ZF, such as the Axiom of Choice, were adopted for extrinsic reasons such as "usefulness," as well as for intrinsic reasons like "intuitiveness." New axioms, from which the CH might be settled, could well be adopted for similar reasons.

This is a good place to leave the matter, as from here the "truth" of the CH becomes more an act of faith than an act of mathematics.

SUMMARY

The Continuum Hypothesis: Does there exist a level of infinity greater than the natural numbers and less than the reals?

Answer: This question is independent of the axiomatic framework customarily used in mathematical argumentation.

CHAPTER FOUR

THE KEPLER CONJECTURE

HILBERT—AGAIN!

In 1976, the American Mathematical Society sponsored a symposium devoted to a review and status report on the famous list of 23 problems that David Hilbert announced as being important for the development of mathematics at the 1900 International Mathematical Congress in Paris. The 18th Problem on Hilbert's list focused on discrete geometry, the third part of which he stated as:

> How can one arrange most densely in space an infinite number of equal solids of given form, e.g., spheres with given radii . . . , that is how can one so fit them together that the ratio of the filled to the unfilled space may be as great as possible?

Commenting on this third part at the AMS symposium, topologist John Milnor wrote:

> For 2-dimensional disks this problem has been solved by Thue and Fejes Tóth, who showed that the expected hexagonal (or honeycomb) packing of circular disks in the plane is the densest possible. However, the corresponding problem in 3 dimensions remains unsolved. This is a scandalous situation since the (presumably) correct answer has been known since the time of Gauss. . . . All that is missing is a proof.

This chapter presents that long-awaited proof—and the surprising way it was obtained—by Thomas Hales of the University of Michigan in August 1998, 22 years after Milnor's challenge.

Interestingly, when Hilbert introduced his list of problems, he said a test of the perfection of a mathematical problem is whether it can be explained to the first person you encounter in the street. Can you imagine talking with your barber about the theory of quadratic reciprocity? Or about the Bieberbach Conjecture with the neighborhood pharmacist? But recently a journalist from Plymouth, New Zealand, put Hilbert's 18th Problem to the test and took it public. The third part of the problem can be restated: Is there a better stacking of oranges than the typical pyramids found at grocery stores? Figure 4.1 shows Thomas Hales stacking up a collection of tennis balls in just this configuration. In these pyramids the oranges fill just over 74 percent of the space. Is there any other packing that would fill a larger fraction? As we'll see in a moment, in 1611 Johannes Kepler conjectured that this was not the case, and that the pyramid (technically: the face-centered cubic) packing is the best that can be done.

The fruit dealers in New Zealand were not impressed by Hales's solution of the Kepler Conjecture. For example, a greengrocer named Allen said, "My dad showed me how to stack oranges when I was about four years old." Told that it took mathematicians over four hundred years to prove that this was the best possible arrangement, Allen was asked how hard it was for him to find the best packing. "You just put one on top of the other," he replied. "It took me about two seconds." In this same flippant vein, shortly after Hales announced his solution he received a call from the Ann Arbor farmers' market. "We need you down here right away," they said. "We can stack the oranges, but we're having trouble with the artichokes."

Hales is a geometer, and it puzzles him why there is such a large gap between intuition, which says it's "obvious" that the best stacking is the pyramid, and proof of the "obvious," which sometimes eludes us for centuries. Let's have a longer look at these matters now within the framework of the long road to the solution of Kepler's deceptively simple conjecture of 1611.

CANNONBALLS AND SNOWFLAKES

Take a large supply of marbles and a box. Drop a bunch of the marbles in the box and shake it up. When you do this, the marbles will arrange themselves into some kind of packing inside the box. Repeating this procedure several times, and measuring the level of marbles in the container, you quickly find that this level is remarkably stable if the shak-

Figure 4.1. Thomas Hales stacking oranges.

ing is performed in a thorough manner, even though it's highly unlikely that the *exact* arrangement of marbles in the box could be identical each time. Nevertheless, the effect is well known in real life, as grocers and customers all believe that a pound of coffee in a can is well-defined.

What we're seeking by such an experiment is the best way to pack spheres into a given volume of space. In particular, we're looking for how to pack the maximum number of spheres into that space. If the fraction of the box taken up by the volume of the marbles is f, then the space between the balls must be $1 - f$. What is f for the densest possible packing? By experiment, we discover that the random packing described above leads to a value of $f \approx 0.64$.

Because of the way they are prepared, random packings are incompressible; squeezing the spheres together doesn't cause any movement. Mathematicians call this a local maximum of the density, which means that small changes in the positions of the spheres causes the density to decrease. But Kepler's problem was to find a *global* maximum of the density, an arrangement in which the density cannot be increased even by big changes in the position of a sphere.

Local maximality has a rather amusing side story, involving the compressibility of sand. If you walk on a beach, does the sand beneath your feet get compressed by your weight when you step on it? It certainly seems that way. But ...

If the sand is in an incompressible random packing, which appears very plausible, then it *cannot* be compressed any further; hence, it cannot be compressed when you step on it. So if any particles of sand move, then the space between them must increase. In 1885 Osborne Reynolds reported to the British Association that, "As the foot presses upon the sand when the falling tide leaves it firm, that portion of it surrounding the foot becomes momentarily dry." He explained that the pressure of the foot actually causes the sand to *dilate*, opening up new gaps for water to flow in from the surrounding sand. This little story shows that sphere packings can be very counterintuitive. Experiments suggest that random arrangements, even though they are incompressible, do not even come close to the densest packing possible.

Suppose we change the experiment a bit. Now instead of shaking up the balls in the box, we place them carefully one by one in a way that is not random but has some repeating pattern. After a few fumbling experiments, you'll soon discover that there is an ordered dense packing that seems particularly efficient. It is exactly the one often seen in fruit stands or the stacking of cannonballs at war memorials. This packing is the one earlier termed the face-centered cubic—or pyramid—packing. It's obtained by placing three marbles on a flat surface with each one touching the other two. The centers of the three marbles then form an equilateral triangle. We then continue packing by adding spheres on the surface so that each new one touches at least two of those already in place. In this way we obtain a layer of marbles in which each touches six others, except for those on the "outside" where we decide to stop. But if we continued on "to infinity," filling up all of space with this packing of marbles, this packing procedure will yield a packing fraction $f = \pi\sqrt{18} \approx 0.74048$—substantially denser than the random packing. What Kepler claimed—and Thomas Hales proved—is that there is no packing of spheres in space that has a higher value of f (although there are many with this same value). Let's take a short look at some of the history of this problem.

In the sixteenth century Sir Walter Raleigh asked his assistant, Thomas Harriot, to develop a formula for the number of cannonballs that were in various stacks on the deck of his ship. Harriot gave Raleigh a formula—but realized that he didn't really know whether the cannonballs were being stacked by the sailors in the most efficient fashion. So he passed the problem along to Johannes Kepler in a piece of correspondence on the theory of atomism written at the end of 1606, asking if Kepler could provide an answer to the *best* way to stack the cannonballs.

Kepler later incorporated his thoughts on the question into a New Year's present to his sponsor, the "illustrious Counsellor at the Court of His Sacred Imperial Majesty, John Matthew Wacker of Wackenfels, Knight Bachelor, ... , Patron of Men of Letters and of Philosophers, my Master and Benefactor." Whew! This little intellectual present was titled *On the Six-Cornered Snowflake.* In its pages, Kepler argues that while no two snowflakes are exactly alike, they all have hexagonal symmetry (they can be rotated through multiples of sixty degrees without changing their appearance). Kepler asked Why? His study of this question led to major advances in the theory of crystals.

A key element of Kepler's treatment of crystalline structures and patterns is the idea of a space-filling solid: a polyhedron (like a cube of a dodecahedron) such that identical copies can be stacked together without leaving any gaps. Kepler then quickly develops this notion of filling up space with copies of a single geometric structure into the mathematical problem of packing identical spherical pellets into the smallest possible space. He finds that in the plane there are two ways to arrange the spheres (circles, in that two-dimensional situation): in a square lattice, like the squares of a floor tiling or checkerboard, and in a hexagonal lattice, like the cells of a honeycomb. These two-dimensional patterns can then be stacked in a variety of ways.

- *"Square" layers:* This stacking puts corresponding spheres vertically above each other, with the centers of the spheres being at the corners of cubes that stack together to fill space. Kepler says, "But this will not be the tightest pack."
- *"Pyramid" packing:* This involves putting the spheres in each layer into the gaps between the four spheres in the layer below. This is the face-centered packing.
- *"Hexagonal packing:* In this stacking, the spheres in each layer also go into the gap between the four spheres below, but each layer has the hexagonal pattern instead of that of a square.

Each of these three stackings is shown in Figures 4.2a–c.

Now, says Kepler, let the spheres in all three arrangements expand. This leads to three different characteristic shapes. For the cubic lattice, the layers become cubes; for the hexagonal lattice, hexagonal prisms; for the face-centered cubic lattice, they become rhombic dodecahedra. Kepler goes on to argue that natural objects like the seeds of pomegranates choose to become rhombic dodecahedra because their seeds can "jiggle around" as they grow. So they will try not to waste any space. This leads to the face-centered cubic lattice, for which he then asserts "the packing will

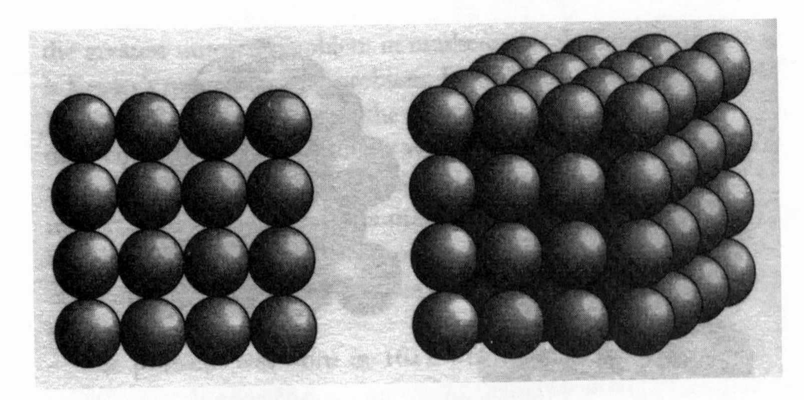

Figure 4.2a. Simple cubic stacking.

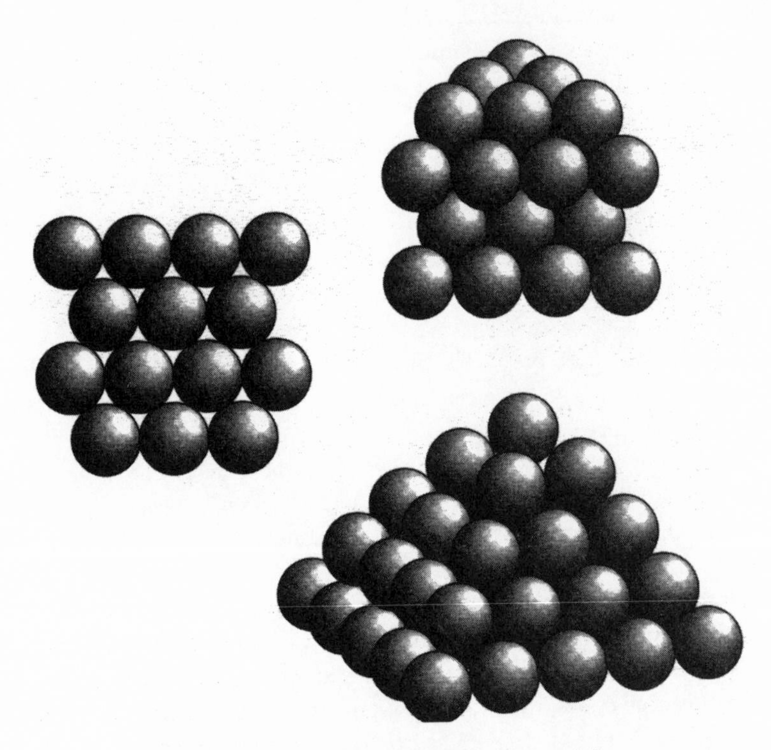

Figure 4.2b. Face-centered cubic stacking.

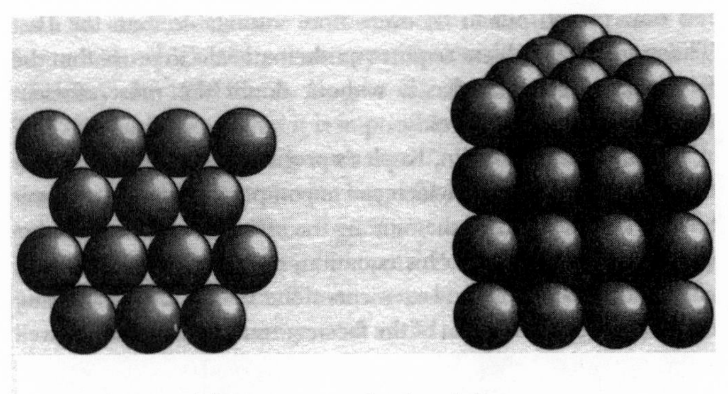

Figure 4.2c. Hexagonal stacking.

be the tightest possible." This seven-word claim stood as a challenge to mathematicians for over 380 years!

That we might take seriously the notion that there may be a special random packing whose density is greater than 0.74048 is best envisaged by looking at what is called the "kissing number" of spheres. This is the number of identical spheres that can be arranged around a central sphere in such a way that all the surrounding ones just touch, or "kiss," the central one. In two dimensions it's easy to see that this number is six, a conclusion that you can easily validate just by taking some pennies from your pocket and laying them out on a table in a hexagonal pattern. But in three dimensions the experiment is by no means so easy to either carry out or see an obvious answer.

The three-dimensional sphere-kissing problem was the subject of a famous dispute in the year 1694 between Isaac Newton and the Scottish astronomer David Gregory. Newton maintained that the kissing number was 12 (the number of spheres found in the face-centered cubic array), while Gregory believed that a 13th sphere could be squeezed in, although he was not able to prove it. His basic idea was that the 12 touching spheres from the face-centered-cubic packing could be rolled around the central one in such a way that the gaps, which are certainly there in the packing, could all be concentrated in one direction, thereby making room for an extra sphere to be inserted to touch the middle one. Only in 1874 was it finally proved that the gap that is accumulated in this fashion is not large enough to accommodate the extra sphere. So the correct answer to this three-dimensional sphere-kissing problem is Newton's number, 12.

Despite this finding, it's possible to start building a cluster with a few spheres in a way that seems to be doing better than the face-centered cubic from a packing density point of view. For example, suppose we start with a "seed" of four spheres packed together so that each touches the other three. When the four spheres are all touching, their centers are at the corners of a regular tetrahedron. This is a figure with four identical triangular faces, essentially a triangular pyramid, each face being an equilateral triangle. Since each sphere in this configuration touches the other three, it must be the densest possible configuration for four spheres in space.

Now suppose we add to this tetrahedral seed other spheres, one at a time, so as to make up a new tetrahedral configuration at each stage. All we need to do is to arrange for each new sphere to be added in such a way that it makes contact with three others. A few experiments with marbles shows that there seems to be no problem in doing this. If we continue in this manner the packing obtained will certainly have the maximum possible density, which is about 0.7796. The trouble is that as we proceed, there always comes a time when the next sphere just cannot be added in the fashion just prescribed. At a certain stage of the growing cluster, the surface acquires a shape that simply does not allow another sphere to be added that touches three existing ones. When this breakdown occurs it is disastrous; large volumes of space are now unavoidably wasted in a manner that reduces the packing density dramatically to values far below that of the face-centered cubic, back toward the experimental limit of around 0.64 for purely random packings.

What this shows, in effect, is that space cannot be completely filled up with regular tetrahedra in the manner that it can be packed with cubes. With respect to sphere packings, a "greedy algorithm" of trying to make the next step the most efficient takes immediate gains at the expense of leaving the configuration in a state that will lead to major difficulties at some time in the future of the packing process. In August 1998, Thomas Hales, together with his graduate student, Samuel P. Ferguson, proved the following theorem that finally closed out the lingering question of whether an especially clever packing of spheres could do better than the face-centered cubic:

Kepler Conjecture. No packing of balls of the same radius in three dimensions has density greater than the face-centered cubic packing.

Let's look briefly at some of the approaches and partial results leading up to this long-awaited—and long-believed—theorem.

PACKING THEM IN

It was Gauss, the Prince of Mathematicians, who was the first person to prove anything useful about the Kepler Conjecture. In 1831, he imposed the condition that centers of the spheres of a packing should be aligned along the points of a lattice. To see what this means, envision an infinite checkerboard and place a sphere in each square so that its center rests above the center of the square. It's easy to see that in the best possible case two spheres will touch each other, and then the lattice constraint forces the spheres to touch along long parallel lines of spheres, much like a row of beads on a necklace. In the best case it will also be true that two of the long parallel lines will touch. Again, the lattice constraint forces the spheres to be laid out in identical parallel sheets. The centers of four spheres in a sheet form a parallelogram, as shown in Figure 4.3.

Now two parallel sheets should be set one atop the other so that the sheets are as close together as possible. A sphere D of the top sheet is set in the pocket between the three spheres A, B, C in the sheet below (the one shown in Figure 4.3) so that it touches all three. The triangle ABD formed by the centers is then an equilateral triangle.

If we now change our point of view, considering all the spheres as arranged in sheets parallel to ABD. In each of those sheets the centers of the spheres repeat the pattern of the equilateral triangle ABD. The spheres of one sheet should then be nestled in the pockets of the sheet before so that each sphere rests on the three below it. The lattice described by this packing is then the face-centered cubic, which Gauss showed is the best possible packing—under the constraint that the centers of the spheres sit on the points of the lattice.

In the world of number theorists, lattice packings soon took off and today they have many applications in communication and coding theory. Let's take a quick look at how this issue of sphere packing enters in a big way into the problem of coding information effectively.

Error-Correcting Codes

When data is stored or transmitted, errors might occur. Coding theory aims at detecting, and if possible, correcting those errors. Information transmitted from a satellite taking pictures, say, of Jupiter may be com-

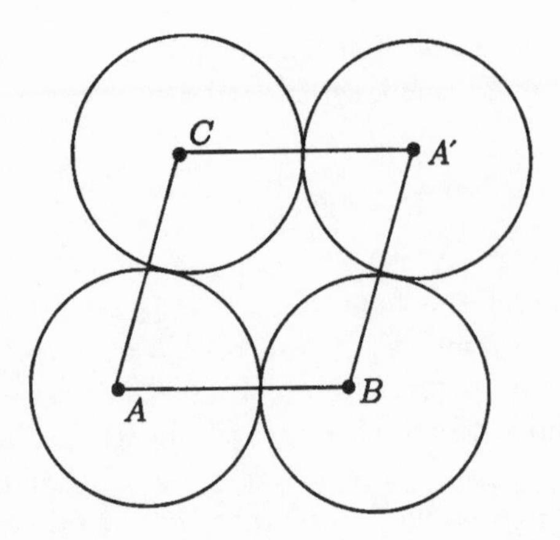

Figure 4.3. Alignment of four spheres in an optimal lattice packing.

promised in many ways and it makes more economic sense to anticipate such errors and try to correct them than to launch another satellite if the pictures are too fuzzy.

When transmitting data through any type of communication channel, the basic medium is a signal that has two different states, call them 0 and 1. Every message is then built up as a string of these symbols, a so-called "bit string." Any incorrect character in the string can cause an error, from a typographic mistake to the termination of a computer program.

For simplicity, suppose our model of communication involves a sender encoding data in a message, also called a *codeword.* This codeword is transmitted over a channel to a receiver. Coding theory deals with what are called "random errors" in the codeword, which are simply alterations in the symbols, a 1 read as a 0 or vice versa. The theory does not consider symbols that are omitted, nor those that are added, and it does not consider situations in which someone is deliberately trying to disrupt the message (disinformation or jamming).

What does it mean for a code to be "error-correcting?" Since the message is always a string of binary digits, we say that the code word is a vector in the space $V = \{0, 1\}^n$. Suppose we transmit a message c and the receiver decodes a message x. Then if x does not equal c, an error has occurred. If there are a "small" number of errors, x and c will be "close." To formalize this notion of "closeness" we need a metric measuring the distance between two vectors in V. The simplest such metric is the so-called *Hamming metric,* which says simply that the distance between two elements of V is just the number of positions at which they differ.

It's easy enough to check that this definition does indeed constitute a metric, which in turn enables us to consider V as a metric space.

Since V is a metric space, we can define a "sphere" in this space as follows: If r is a nonnegative integer and v is a fixed element of V, then the sphere of radius r in V centered at v is all vectors x in V such that the distance between x and v is less than or equal to r.

If all of the vectors (code words) in our code are vectors that are the centers of non-overlapping spheres, then it will be possible to build an error-correcting code. This is because if a code word that is received falls within the radius of one of these code words, then it is simply interpreted as that vector.

We can formalize this intuitive insight as follows. Suppose t is a positive integer. Then a subset C of the vectors of V is called a *t-error-correcting code* if the distance between any two elements v, w of C is at least $2t+1$. In terms of code words, this says that the minimum distance between code words is at least $2t + 1$. Since the spheres constituting the code words are disjoint, for any vector in V, there will be at most one code word within a distance $2t + 1$.

Practical considerations dictate the size of t. On the one hand, if you think the frequency with which more than t errors will occur is negligible, then you build a t-error-correcting code. In this case, a vector with more than t errors cannot be corrected. Thus, the larger t is, the better your chances of reconstructing the original message. However, the larger t is, the "bigger" the nonoverlapping spheres must be, thus adding to the cost of transmission. What is the same thing, if t is large, so is n, the dimension of the space of code words.

This is the connection between coding theory and sphere-packing. In fact, this shows why it is of crucial importance to understand the sphere-packing problem in very high dimensions, since we need to know the maximum number of nonoverlapping spheres that can be squeezed into an n-dimensional space. Luckily, for coding theory the lattice-packing version of the problem is all we need. But we need it in very high dimensions. Interested readers can find more information about sphere-packing and coding in the material cited in the chapter references. Now let's return to the Kepler Conjecture.

Kepler's problem of *non*-lattice packings remained untouched for another 60 years until Axel Thue outlined a solution to the two-dimensional version of Kepler's problem, showing that the densest packing of equal circles in the plane—regular or not—is the hexagonal lattice.

If we inscribe a disk in each hexagon in the regular hexagonal tiling of the plane, the density of the packing is $\pi/\sqrt{12} \approx 0.9069$. Thue's theorem, published in 1890, asserts that this is the densest packing possible.

These methods for the lattice packing problem in the plane offer no help in dealing with the three-dimensional problem, as they make essential use of the flat, two-dimensional structure of the plane. So there remains the possibility of a very efficient *non-lattice* packing. That this is a live possibility is shown by the fact that there exist such non-lattice packings that are *exactly* as dense as the face-centered cubic lattice. Here's how to construct one. Start with a layer of spheres that is a hexagonal lattice in the plane. Now add a second layer whose spheres fall between those of the first layer. We now place the third layer so that its spheres sit directly above the corresponding spheres in the first. Repeating this process for subsequent layers leads to a non-lattice pattern. It's clear from the way the layers fit together that the density is the same for this non-lattice packing as it is for the face-centered cubic.

Moreover, it was shown by W. Barlow in 1883 that we can continue to put on hexagonal layers placed randomly between the two possibilities of nestling the spheres into the pockets of the preceding layers or setting them on top of the spheres of the next-most recent layer. This kind of packing would be random in the vertical direction, but regular in the two horizontal directions. And the density of each of these random packings is exactly the same as that of the face-centered cubic.

All this suggests that it is not at all clear that the best packing—the one with the highest density—must be a lattice packing. In 1958 C. A. Rogers proved that the maximum density for any packing in space is at most 0.7797, which is only slightly greater than the 0.7404 of the face-centered cubic packing. So any packing better than the one conjectured by Kepler could only squeeze in at most an extra slice of a sphere, on the average, and would have to be very cleverly constructed to improve on the face-centered cubic.

In order to take things further, some new ideas were needed.

BACKING AWAY FROM INFINITY

In the full sphere-packing problem, the position of every sphere becomes a variable—and it takes infinitely many spheres to fill up all of three-dimensional space. No computer, however powerful, can solve unaided a problem with an infinite number of variables, since it would require the storage of an infinite amount of data to do so.

The trick is then to reduce the problem somehow to a finite one—to divide up space into small regions and analyze each one separately. If the face-centered cubic pattern could be proved as good as any challenger in every single region, then it would be the best in the entire space, too. Sad to say, this is a hard task, rather analogous to requiring the champion to win each and every round in a boxing match. As we've seen above, early attempts to solve the Kepler Conjecture this way failed.

But in 1953, the Hungarian mathematician Lászlo Fejes Tóth suggested defining a region, or "cell," as the set of points that are closer to one particular sphere's center than any other. Each such cell, called a *Voronoi cell,* contains a spherical "nucleus." In the face-centered cubic packing, every Voronoi cell is a rhombic dodecahedron, whose nucleus fills up 74 percent of the cell. Such a Voronoi cell for the face-centered cubic packing is shown in Figure 4.4. But some less-regular packings can contain denser Voronoi cells—for example, regular dodecahedra, whose nuclei occupy 75 percent of the volume. Therefore these packings would beat the face-centered cubic pattern in some regions.

By introducing these Voronoi cells, Tóth reduced the Kepler Conjecture to a finite—but impossibly large—calculation. In his words

> Thus it seems that the problem [the Kepler Conjecture] can be reduced to the determination of the minimum of a function of a finite number of variables.... Mindful of the rapid development of our computers, it is imaginable that the minimum may be determined with great exactitude.

Let's take a longer look at the strategy that Tóth had in mind, and that Thomas Hales modified somewhat in his final assault on the problem.

In his work, Tóth developed a method in which analyzing a cluster of 50 spheres would be sufficient to prove the Kepler Conjecture. The calculation Tóth presented involves looking at all the positions of the 50 spheres within the cluster. For each set of positions, you can then calculate a packing efficiency. Because the position of each sphere is defined by three spatial coordinates, the equation you need to calculate the packing efficiency consists of 150 variables. The task is to find the maximum value of this function. If the maximum efficiency turns out to be 0.7404, the efficiency of the face-centered cubic packing, then the conjecture would be proved.

One picture shows the entire method. There is a nonlinear function $f(x_1, x_2, \ldots, x_{150})$ that we have to maximize. That is, we want to find values of the variables $x_1, x_2, \ldots, x_{150}$ that make the value of f as large as possible. Part of the difficulty is that the number of variables is very

Figure 4.4. A Voronoi cell for the face-centered cubic packing.

large, 150. Moreover, the function f has no nice analytic structure like convexity or linearity that would make the problem easy to solve.

In calculus, we learn that the way to find the maximum of a function is to compute where the function's first derivative vanishes. It would foolhardy to try this approach for a problem of this nature. There are too many variables and the problem of finding the zeros of the first derivative is at least as difficult as finding the maximum of the original function. We have to do something else.

Instead, we plot the efficiency on a "graph" of 150 variables, building a hilly landscape in 150-dimensional hyperspace. This is just like looking for the peak of a curve—except the curve is now an impossible-to-visualize surface in a space of 150 coordinate directions. All we need then is a method to find the height of the tallest hill. To do this, draw a number of hyperplanes that lie above the graph of the function f, a kind of "roof" over the landscape of f. We then ask for the highest point X that lies beneath all the hyperplanes. This point is always at least as high as the maximum of the original function f. The trick is in lowering the roof until it just touches the tallest hill. Actually, since what we're mostly interested in is a bound on the density of packing, it usually suffices to use this linear approximation to the true nonlinear maximum, and not find the actual maximum itself.

Since X can be found by strictly *linear* operations, linear programming techniques can be used to find X. And linear programming problems with thousands of variables—not just the 150 or so involved in solving the Kepler Conjecture—are solved every day.

The fly in the ointment, however, is that we cannot use just any old hyperplane that lies above the graph of f. We are only allowed to use "elementary hyperplanes." These are hyperplanes that we can rigorously prove lie above the graph of our function f. But as we are working in very high dimensional space, such hyperplanes are very rare. Nevertheless, Hales and his colleagues found enough of them to solve Kepler's Conjecture. Let's see how this magic works.

HAIL HALES

Wu-Yi Hsiang is a professor of mathematics at Berkeley. In the spring of 1990, he agreed to create a new course on a totally outdated branch of mathematics—classical geometry. To motivate the course, Hsiang took a hard look at the most difficult problem in the area he could find, the Kepler Conjecture. Shortly thereafter, he claimed a solution. As he said at the time, "I got hooked on it. The more I thought about this problem, the more beautiful it appeared."

Hsiang's "proof" was also in the classical vein, eschewing all of the machinery developed by mathematicians over the past decades. It is stated entirely in the language of spherical geometry, calculus, vector analysis and other tools in use since the nineteenth century. Hsiang's view is simply that we haven't sufficiently appreciated spherical geometry.

The central idea underlying his approach to the problem is the so-called "local density conjecture." Envision each sphere of a packing as swelling at a uniform rate until it touches its neighbors along flat interfaces. Then each sphere is surrounded by a polyhedron, its local cell. This is simply the Voronoi cell we discussed earlier. The *local density* of any particular sphere is then the ratio of its volume to that of its local cell. If any upper bound for the local density can be found, that bound will automatically be an upper bound for the density of the entire packing.

Hsiang thus started with trying to prove the local density conjecture. As he stated things, "The central issue is how to achieve effective lower bound estimations of the volume of a local cell or that of a cluster of local cells." He found a new method for estimating the volume of local cells by cutting them up. But he did not cut them into cones as had been the standard approach of previous researchers. Rather, he used *double cones,* which he claimed exploited the spherical geometry of local cells more effectively. Using a notion of "semiglobal" density, he then stepped forward from the local version of the problem that involves examination of just a small, finite number of cells to the overall global problem involving an infinite number. This strategy was analogous to allowing the boxing champion to lose a round—but he would then have to win the round after and the one before.

But after more than a hundred pages of esoteric geometry, cracks started to emerge in Hsiang's "proof," and the general consensus of the mathematical community was echoed by Tóth's son, Gabor, when he wrote in *Mathematical Reviews* that "The greater part of the work has yet to be done." Enter Thomas Hales.

Hale's winning strategy involves following the same route to salvation via the local density conjecture—but with a twist. He employs not only the Voronoi cells, but also a second way of chopping up space called the *Delaunay decomposition.* Using a subtle blend of these two ways of creating local cellular clusters, Hales arrives at a finite optimization problem that is computationally feasible. While the details belong to the technical literature, here's the overall approach.

Take any sphere packing and connect the centers of spheres that come within a distance of 2.51 of each other. Did you say 2.51? Why this peculiar number? Well, in Hales's method there are a lot of rather arbitrary sounding constants like this that appear. The value 2.51 is just one of them. The only justification for this particular choice is that after doing numerous experiments, it seemed as good a choice as any and some cutoff had to be made. The lines connecting the center of any two such spheres form a graph. If we fix one sphere center and look at just the edges between sphere centers that are within distance 2.51 of this fixed center, we get a finite graph. For example, if you start with the face-centered cubic packing, the resulting graph is one with 24 edges. This object is the principal object of Hales's investigation of Kepler's Conjecture.

There is a decomposition of space that is dual to the Voronoi decomposition considered earlier. Every face of a Voronoi cell consists of points equidistant from two different sphere centers. Connect every such pair of sphere centers by an edge. These edges divide space into simplices called Delaunay simplices. Since each such simplex has four vertices, it constitutes a tetrahedron, the collection of all such objects forming a full decomposition of three-dimensional space. A *Delaunay star* is defined to be the union of all the Delaunay simplices around a common vertex. The Delaunay stars overlap. In fact, each Delaunay star lies in four Delaunay simplices, one for each of the four vertices of the simplex. It's now possible to reformulate the Kepler Conjecture in terms of these Delaunay stars.

We can define a function on the set of Delaunay stars, called the *score* of the star. In practice, it's easier to draw graphs than stars, so we usually think of the scoring function as being defined on the graph equivalent to the star rather than on the star itself. The score is the sum of the scores for each of the faces of the planar graph. A triangle scores at most 1 point, with exactly 1 point given only for the triangle coming from a tetrahedron. Any other face gets a negative score.

For example, there are 8 triangles in the graph of the face-centered cubic packing. They score exactly 8 points, 1 point per triangle. In terms of scores, the Kepler Conjecture is true if the score of every Delaunay star is at most 8 points. While this fact is relatively easy to prove, it would take us too far afield to give the proof here. The interested reader can consult the work cited in the references for the details. What Hales saw is that a hybrid of the Voronoi and Delaunay decompositions would work better at relating the scoring function to a decomposition of space, and would lead to a proof that the score of every Delaunay star is at most 8 points. Note that this must be shown for *every* possible Delaunay star, and many different stars lead to the same graph.

With the scoring function in hand, in 1994 Hales outlined the following five-step program to prove the Kepler Conjecture:

1. Treat graphs associated to a sphere that have only triangular faces. These types of graphs are exemplified by the one shown in Figure 4.5.
2. Show that the face-centered cubic and the hexagonal packings are local maxima among all packings, in the sense that they have a higher score than any Delaunay star with the same graph.
3. Treat maps that contain only triangular and quadrilateral faces. An example of this type of graph with a quadrilateral face is shown in Figure 4.6.
4. Treat maps that contain something other than a triangular or quadrilateral face.
5. Treat pentagonal prisms, graphs in which the centers of the spheres form a five-sided shape. (Recall that the pentagonal prism includes Delaunay stars that have higher scores than the hexagonal packing when a pure Delaunay decomposition of space is employed. So in this case the estimates of the scoring function value must be done with great delicacy and precision.)

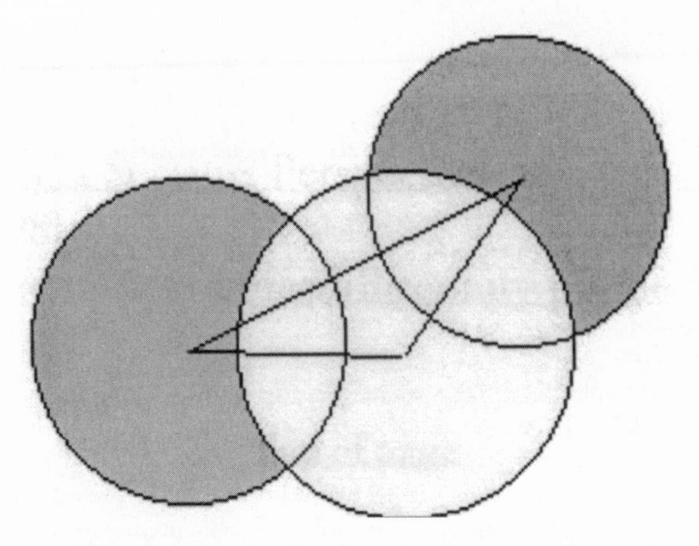

Figure 4.5. Structure of graphs with only triangular faces.

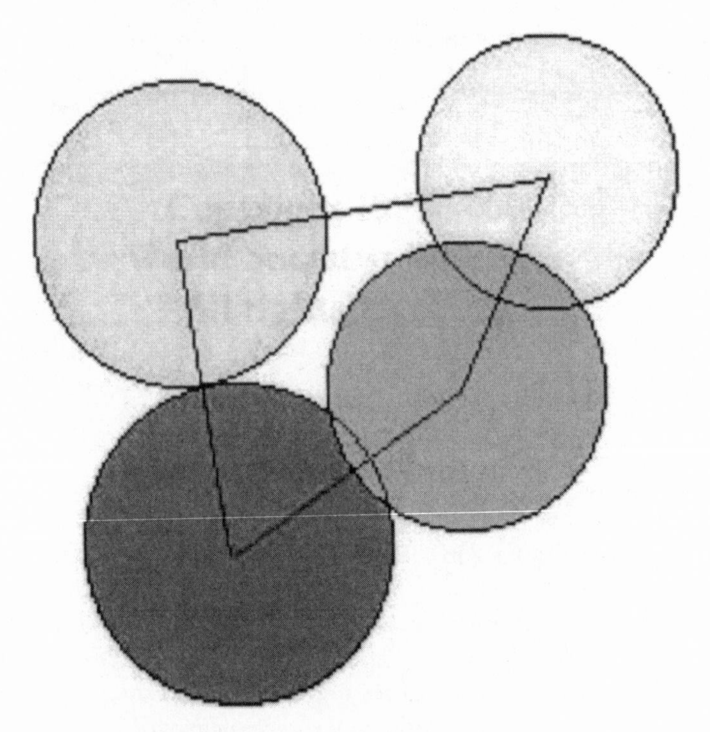

Figure 4.6. A graph with a quadrilateral face.

PROOFS AND COMPUTERS—REDUX

Generally speaking, there are three grades of proof in mathematics. The first, or highest quality type of proof, is one that incorporates *why* and *how* the result is true, not simply that it is so. Such proofs may be quite simple, as for example Euclid's proof that any triangle with equal base angles must be isosceles. A more complicated proof of this type is Gauss's demonstration that the only polygons constructible by ruler-and-compass are those whose sides are based on the Fermat primes.

Second-grade proofs content themselves with showing *that* their conclusion is true, by relying on the law of the excluded middle. Thus, they assume that the conclusion they want to demonstrate is false and then derive a contradiction from this assumption. In polite company, these are often termed "nonconstructive proofs," since they lack the how and why. Euclid's proof of the infinity of prime numbers is of this type. To be a proof of the first order, Euclid would have had to produce a formula for constructing primes, and then show that this method would generate prime numbers ad infinitum—literally. Instead he assumed a finite number of primes and derived from this a contradiction, establishing their infinitude instead.

Finally, there is the third order, or lowest grade, of proof. In these situations, the idea of proof degenerates into mere verification, in which a (usually) large number of cases are considered separately and verified, one by one, very often by a computer. The work by Haken and Appel solving the Four-Color Problem is of this type, as is the proof just outlined of the Kepler Conjecture.

Thomas Hales and his graduate student, Samuel Ferguson, showed that there are about 5000 graphs that are relevant to the proof of the Kepler Conjecture. Using general linear programming methods to estimate bounds on the scoring functions, problems of 150 variables with up to 2,000 side constraints had to be solved—an "easy problem" for today's linear programming methods to handle. Hales and Ferguson reduced this list to around 100 cases, which were then eliminated, one by one, to provide the final proof of Kepler's Conjecture. The computer programs required for these calculations required more than three gigabytes of storage (3,000 million bytes), an amount that would have been prohibitive just a few decades ago. But the volume of calculations involved—and the central role they played in the proof—again raises the same issues regarding computer proofs that were faced in the Haken-Appel solution to the Four-Color Problem. As Hales stated,

> As this project has progressed, the computer has replaced conventional mathematical arguments more and more, until now nearly every aspect of the proof relies on computer verifications. Many assertions in these papers [the proof] are results of computer calculations.

Here are some of the ways that Hales and Ferguson used the computer in their work:

- *Interval arithmetic:* Interval arithmetic allows computation using confidence intervals for the quantities involved in the computation rather than computation with single numbers. Thus it is especially well-suited for calculating upper and lower bounds of various quantities in the work on Kepler's Conjecture. Inequalities in a small number of variables were verified by Hales and Ferguson using these computational techniques.
- *Combinatorics:* Computer programs classified all the graphs relevant to the Kepler problem.
- *Linear programming bounds:* Many of the nonlinear optimization problems for the scores of decomposition stars are replaced by linear problems whose solution dominates the original score. These are solved by linear programming methods by computer. As noted, a typical problem involves around 150 variables and between 1,000 and 2,000 constraints. Nearly 100,000 such problems enter into the proof.
- *Branch-and-bound methods:* Sometimes the linear programming methods do not give sufficiently tight bounds, in which case they are augmented by branch-and-bound methods for nonlinear, global optimization.
- *Numerical optimization:* Many nonlinear optimization and symbolic mathematics packages were used at various stages in the proof.
- *Organization of the output:* It is a far from trivial task to organize the three gigabytes of computer code and data that enter into the proof.

With all these various uses of the computer, it's clear that the proof offered by Hales and Ferguson cannot be surveyed by any single mathematician or group of mathematicians. To manually go through this code, line by line, would be impossible. Moreover, computers have known failure rates, implying that a certain number of errors were almost surely made during the course of these myriad calculations.

The second of these objections is easily answered by noting that computers are certainly less prone to error than their human counterparts.

As for the first argument, while it is not possible for humans to replicate execution of the programs, it is possible for the method of their execution to be analyzed—as long as the computer code is published as part of the proof itself. Hales has done precisely this, so that interested observers can download the programs from the Internet and check them for themselves.

So just as with the Four-Color Problem, there is very little doubt of the validity of Hales's solution to the Kepler Conjecture, even though some unreconstructed mathematical purists continue to seek a solution "by hand," so to speak. Now let's have a quick look at a variety of other areas in which sphere-packing or its close mathematical relatives play an important role.

BEES AND SHAVING CREAM

In an essay on "the sagacity of bees," Pappus of Alexandria observed in the fourth century how bees, possessing a divine sense of symmetry, had as their mission the fashioning of honeycombs without any cracks through which that wonderful nectar known as honey could be lost. Pappus focused his mathematical analysis of the bees's problem on their hexagonal arrangment of cells, investigating the amount of wax it takes to contruct a comb.

As honeybees gorge themselves on honey, they slowly excrete slivers of wax, each fleck about the size of a pinhead. Other workers harvest these tiny wax scales, then carefully position and mold them to assemble a vertical comb of six-sided, or hexagonal, cells. The bees cluster in large numbers, maintaining a hive temperature of 35 degrees centigrade, which keeps the wax firm but malleable during cell construction.

Observers throughout history have marveled at the hexagonal pattern of the honeybee's elaborate storage system. In the 19th century, Charles Darwin described the honeycomb as a masterpiece of engineering that is "absolutely perfect in economizing labor and wax."

Biologists have assumed that bees minimize the amount of wax they use to build their combs. But is a grid made up of regular hexagons (those with sides of equal length) really the best possible? What if the walls were curved instead of being flat? Or if they bulged outward or inward in some particularly tricky way?

While the honeycombs are three-dimensional structures, each cell is uniform in the direction perpendicular to its base. Hence, its hexagonal cross-section matters more than any other factor in determining the amount of wax it takes to construct a comb. Thus, the mathematician's

Honeycomb Conjecture concerns the shape of tiles of a fixed area, say size 1, that would (a) completely cover the plane, and (b) do so with the smallest possible perimeter. The Honeycomb Conjecture states that regular hexagons are the best shape (these are six-sided polygons, all of whose sides are of equal length). In that case, the best possible tiling would look like the one depicted in Figure 4.7.

The mathematicians of ancient Greece asked what choices bees might have made if they wanted to divide a flat surface into identical, equal-sided cells. Only three regular polygons fill up the plane without gaps: equilateral triangles, squares, and hexagons. Other shapes like pentagons and octagons will not fit together without leaving spaces between the cells.

The Greeks claimed that if the same quantity of wax were used for constructing a single three-dimensional version of each of the three candidate figures, the hexagonal cell would hold more honey than a triangular or square cell. Or, what is the same, the perimeter of the hexagonal cell enclosing a given area is less than that of a square or trianglar cell enclosing that same area.

It's fairly simple to prove that a regular hexagon has a smaller perimeter than any other six-sided figure of the same area. Moreover, polygons with more sides than the hexagon, such as regular octagons, do better, and polygons with fewer sides, such as squares, do worse. But, of course, octagons are regular polygons that don't fill the plane entirely, so they are noncandidates to solve the problem.

In 1943, Hungarian mathematician L. Fejes Tóth, whom we met earlier in connection with sphere-packing, proved the Honeycomb Conjecture for the special case of filling the plane with any mixture of straight-sided polygons. Tóth established that the average number of sides per cell in a plane-filling pattern is at most six. Moreover, the advantage of having some polygons with more than six sides is less than the disadvantage of having some polygons with fewer sides. So under the constraint that the tiles must have straight sides, the regular hexagonal grid is the least-perimeter way to fill up the plane.

But there are other possibilities for arrays of cells. There is no a priori reason why the cells must all have equal sides or identical shapes and sizes. What about some crazy quilt of different types of polygons or cells with curved rather than straight sides? Sifting through the various alternatives is a major mathematical task, one that was finally completed in June 1999.

In that month, the very same Thomas Hales who solved the Kepler problem also announced a proof of the Honeycomb Conjecture in its

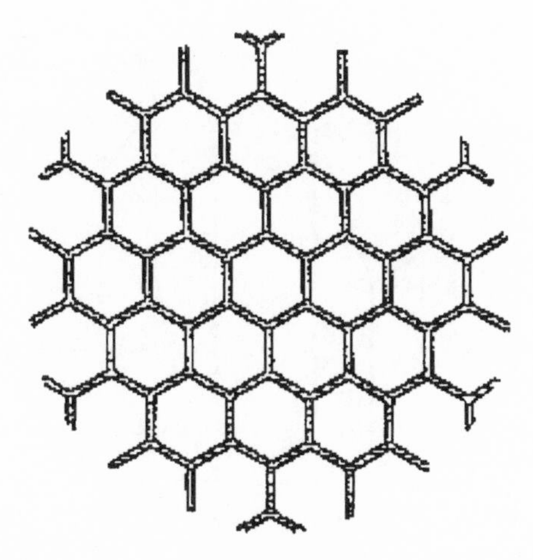

Figure 4.7. Tiling of the plane with regular hexagons.

full glory, where the cells of the tiling are not constrained to be straight-sided polygons. He showed that while having sides that bulge out helps, bulging in hurts. And, of course, a figure cannot have sides that bulge out everywhere; if some bulge out, then others must bulge in to compensate. The question is whether there is a tiling in which the bulging-out does more good than the bulging-in does harm.

The key element in Hales's proof is a "isoperimetric" estimate for perimeter in terms of area, with penalties for bulging out or for more than six sides. Of course, since the plane is infinite we can really only talk about limiting averages of perimeter per region. The proof begins with a reduction to finite clusters (shades of the Kepler Conjecture!), which requires assuming that the regions are connected. Following up this idea, Hales concluded that the advantage of bulging out is less than the disadvantage of bulging in. Therefore, straight-sided polygons are best, and regular hexagons are the best of the best.

But a honeycomb consists of more than just a vertical, hexagonal grid. It is actually built of two layers of cells placed back to back. The cells themselves are tilted upward at an angle of about 13 degrees from the horizontal—just enough to prevent stored honey from dripping out.

Instead of a flat bottom, each cell ends in three four-sided, diamond-shaped panels, meeting in a point. The cells of the two layers are offset so the center of a chamber on one side is the corner of three adjacent cells on the other side. The overall arrangement is shown in Figure 4.8. The structure on the right is an end cap consisting of two hexagons and two

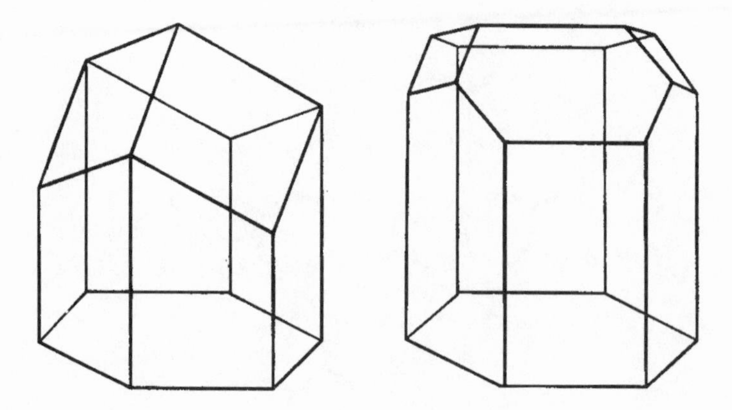

Figure 4.8. The possible structures for the closed ends of a honeycomb cell.

squares. In 1964 Tóth showed that this requires just a bit less wax than the one on the left, which is the one the honeybees actually use. The difference, though, is very small. He says, "By building such cells, the bees would save per cell less than 0.35 percent of the area of an opening (and a much smaller percentage of the surface area of a cell)."

A few years ago, Denis Weaire of Trinity College Dublin and a colleague, Robert Phelan, experimented with a liquid-air foam to test Tóth's mathematical model. They pumped equal-sized bubbles, about 2 mm in diameter, of a detergent solution between two glass plates to generate a double layer—just like the double layer that actually forms a honeycomb. The two layers of trapped bubbles formed hexagonal patterns at the glass plates. The interface between the two layers adopted Tóth's structure.

When Weaire and Phelan thickened the bubble walls by adding more liquid, though, they unexpectedly found an abrupt transition. When the walls reached a particular thickness, the interface suddenly switched to the configuration of a honeycomb. The switch also occurred in the reverse direction as liquid was removed.

Thus, honeybees may well have found the optimal design solution for the thicker wax walls of their honeycomb cells.

This discussion has been focused on two-dimensional honeycombs, in which we ask about planar arrangements of cells of different shapes. What about the real-world question of how to fill up three-dimensional space? What kind of solid objects completely fill space and have the lowest possible surface area? This is the real honeycomb problem.

We saw in Figure 4.8 that the cells of the honeycomb are sealed by a six-sided prism whose faces are three rhombi. Thus, the cell becomes a rhombic dodecahedron. In the eighteenth century, mathematicians made extensive use of the isoperimetric property of the honeycomb, and came to believe that it is the most efficient design possible. But as we also just saw, they were wrong; Fejes Tóth discovered a more economical form shown in Figure 4.8, although the *most* economical shape has not yet been found.

Suppose we tile space with hollow rhombic dodecahedra and imagine that each has walls made of flexible soap film. We then have an example of a foam, just the type of thing that comes out of a can of shaving cream when you push the button. Lord Kelvin posed the question: How can space be divided into cavities of equal volume so as to minimize the surface area of the boundary? The rhombic dodecahedral example is very far from optimal. Kelvin's attempt at a solution involves using truncated octahedra to fill space (see Figure 4.9). The cross sections of these truncated octahedra are regular octagons, which tile the plane except for square holes. Each truncated octahedron contains square plugs, so that the next layer of octahedra plugs the square holes of the previous layer.

Kelvin found that by warping the faces of the truncated octahedra ever so slightly, he could obtain a foam with smaller surface area than the cells of the truncated octahedra. While he thought this was the minimal boundary area solution to the problem, Kelvin was unable to give a proof of this "fact." As it turned out, he was wrong.

In 1994 Weaire and Phelan produced a foam with cavities of equal volume having a considerably smaller surface area than the Kelvin foam. The Weaire-Phelan foam contains two different types of cells, one with 14 sides, the other with 12 (see Figure 4.10). To obtain their foam, imagine a cube with a small ball at each vertex. Form the Voronoi cells around each ball, adjusted so that they all have the same volume. If we then warp the faces of this configuration ever so slightly, the Weaire-Phelan foam results.

The Kepler Conjecture and the Kelvin Problem are both special cases of a more general foam problem. Weaire and Phelan ask us to imagine that the soapy film walls have a measurable thickness. We can then interpolate between the Kepler and Kelvin problems using a parameter w, which measures the "wetness" of the film. This parameter gives the fraction of space filled by the thick film walls. If the foam is perfectly dry, then $w = 0$ and the film walls are surfaces. The Kelvin Problem asks for the most efficient design. When the foam becomes sufficiently wet, w is close to 1 and the cells of the foam can be independently molded. The isoperimetric inequality states that they minimize surface area by forming

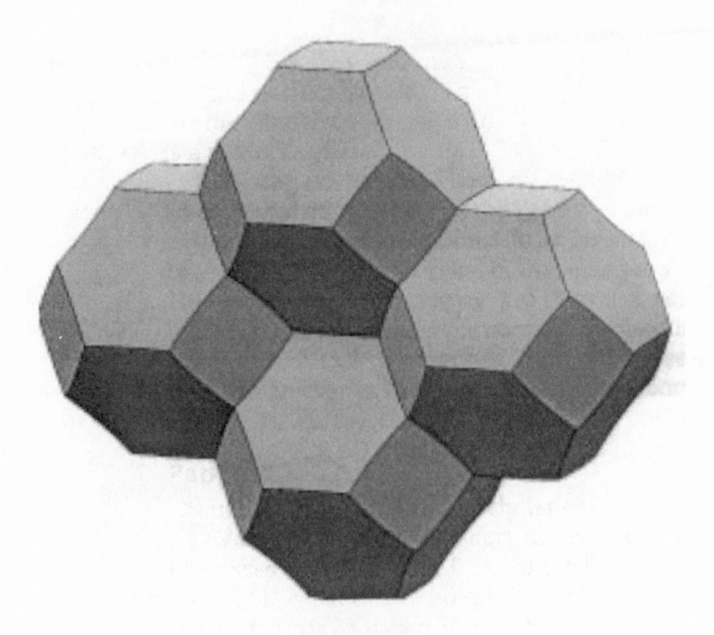

Figure 4.9. Kelvin's partitioning of space using truncated octahedra.

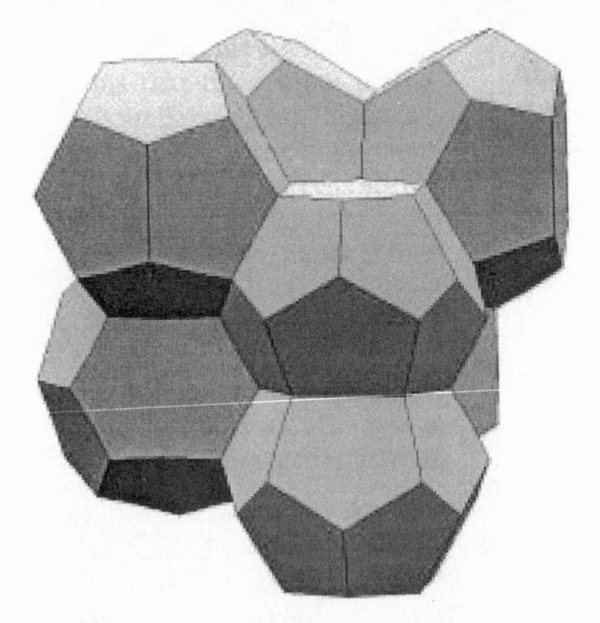

Figure 4.10. The Weaire-Phelan counterexample to the Kelvin foam.

into perfect spheres. The Kepler Problem then asks for the smallest value of w for which every cell is a perfect sphere.

This is the end of our story. The honeycomb problem is preparation for the real challenge, which is the Kelvin Problem. What is the most efficient packing of space into equal volumes? Is the Weaire-Phelan foam the best way to do it? At present, no one knows. On this tantalizing note, we summarize the Kepler Conjecture.

SUMMARY

The Kepler Conjecture: The density of a packing of equal-sized spheres in three dimensions is never greater than $\pi/\sqrt{18} \approx 0.74048\ldots$.

Answer: This is indeed the case. There exists no denser packing of spheres.

CHAPTER FIVE

FERMAT'S LAST THEOREM

REALLY POPULAR MATH

July 28, 1993, San Francisco, California. Five-dollar tickets are scalped for over $25, as 1,000 people fill a hall for an evening of public lectures expounding on the solution of a 350-year-old mathematical problem. Did you say mathematics? What kind of mathematical problem could engender such public interest? Well, perhaps the success of this Fermat Fest, sponsored by the Mathematical Sciences Research Institute in Berkeley, is a bit less surprising when we look at the problem discussed at the event. It was none other than the most famous (previously) unsolved problem in mathematics, the so-called Fermat Conjecture. Moreover, it was a problem so simple that a grade-school child could understand it, yet so impenetrable that literally millions of people worldwide had tried their hand at solving it in the centuries since it was first stated by the famed French jurist and amateur mathematician, Pierre Fermat. As a man at the Fest remarked, "[the conjecture] holds fascination because so many people have tried to prove it." And one woman brought her son, a high-school freshman, who she claimed is something of a math whiz. She stated that she didn't expect to understand much herself, and brought her knitting along to pass the time. But as the evening wore on, the knitting took second place to the lecture podium, as she listened intently to the goings-on. So what is this problem that captured the imagination, time, and energy of so many people?

The Fermat Conjecture is very simple. It asserts that the equation

$$x^n + y^n = z^n, \qquad (*)$$

has no solutions in positive integers x, y and z for any positive integer values of n greater than 2. Of course, if $n = 1$ there are an infinite number of solutions (for example, just choose $x = y - z$ for values of y greater than z.) Similarly, it has been known since the time of the Babylonians that the Pythagorean equation

$$x^2 + y^2 = z^2 \qquad (**)$$

has an infinite number of solutions, such as $x = 3, y = 4, z = 5$ and $x = 5, y = 12, z = 13$ (although it seems that this result was only actually proved by the Greeks many centuries later). Of course, once you have one such "Pythagorean triple" you can get another by simply multiplying all three numbers by any other number you like. For example, if we multiply 3, 4, and 5 by 2, then the new triple $x = 6, y = 8$, and $z = 10$ is easily seen to be another solution of the Pythagorean equation $(**)$.

Fermat's claim was that the process stopped right there: for no value of n greater than 2 is there *any* solution to Eq. $(*)$, let alone an infinite number of them.

No doubt part of the appeal of the Fermat Conjecture is that it's so simple to state and understand. But another, perhaps even greater, part of its allure is the way it was originally stated and how it came to the attention of the mathematical community. So let's briefly go through a bit of mathematical history, by way of getting a glimpse of how a mathematical problem became "famous."

THE PRINCE OF AMATEURS

When Fermat died on January 12, 1665, he was one of the most famous mathematicians in Europe. This is perhaps surprising only because his days were not spent doing mathematics. Rather, he was a lawyer attached to the provincial parliament in Toulouse, France, his hometown. This was a position he obtained at the age of 30, and which he held the remainder of his life.

As a jurist, Fermat needed to maintain rectitude and to remain above suspicion and thus needed to be beyond bribery or other improper influence. So he tended to avoid social engagements and other types of events and encounters that might have compromised his judicial position. Consequently, he had considerable solitude in his daily life, much of his free time was then available to devote to mathematics. But with only a couple of minor exceptions, Fermat avoided publishing any of his work during his lifetime. (It's hard to be optimistic about tenure for this guy!) Nevertheless, he maintained a voluminous correspondence with the greatest mathematicians of his era—Descartes, Pascal, Johann Bernoulli—and made sufficient contributions to the foundations of number theory, geometry, calculus, and probability theory to earn the sobriquet, "The Prince of Amateurs."

The way the Fermat Conjecture, or as it came to be called, *Fermat's Last Theorem,* arose is a twisted and interesting one. When Constantinople fell to the Turks in 1453, scholars fled to the West, bringing the manuscripts of Greek wisdom with them. Among this collection of material was a copy of what had survived of Diophantus's *Arithmetica.* This work remained pretty much unread until 1621, when Claude Bachet published a new edition of the original Greek text, along with a Latin translation, complete with footnotes and commentary. This edition brought Diophantus's work to the attention of European mathematicians—including Fermat.

The greater part of the *Arithmetica* was devoted to the solution in rational numbers of equations in two or more variables having integer coefficients. As discussed in the chapter on Hilbert's Tenth Problem, mathematicians today usually restrict themselves to finding integer solutions to such equations, in which case they are termed "Diophantine equations," in Diophantus's honor.

As he made his way through Bachet's edition of the *Arithmetica,* Fermat often made marginal notes to himself about various approaches and solution methods for the problems he encountered. Five years after Fermat's death, his son Samuel collected all his father's notes and letters for publication. When he came upon the annotated copy of the *Arithmetica,* Samuel decided to publish a new edition, one that would include his father's marginal notes as an appendix. This book was titled *Observations on Diophantus.* The second of these 48 "observations" was a note on Problem 8 of Book II, which asks "Given a number which is a square, write it as the sum of two other squares." We have already seen that this Pythagorean Problem has an infinite number of solutions. In his marginal note on this problem (written in Latin), Fermat states that:

> On the other hand, it is impossible for a cube to be written as a sum of two cubes or a fourth power to be written as a sum of two fourth powers or, in general, for any number which is a power greater than the second to be written as a sum of two like powers. I have a truly marvellous demonstration of this proposition which this margin is too narrow to contain.

This marginal note is what came to be called Fermat's Last Theorem, presumably because of all the many statements of theorems he left behind at his death, this was the last one to be proved. So what Fermat's marginal note says is that if n is a natural number greater than 2, then the equation $(*)$ above has no solution in positive integers.

Did Fermat really have the "truly marvellous demonstration" that he claimed? While the possibility of this is a large part of the charm of the problem, it seems unlikely. The two special cases for $n = 3$ and $n = 4$ are stated elsewhere in his work, whereas mention of the Last Theorem is confined to that one brief margin note. Most likely, he saw how to prove the result for these two cases, and thought that his method generalized to the case of all n greater than 2, but later discovered that it did not. Since he never intended that the marginal notes be published, he had no need to go back and amend them to reflect this disappointing knowledge. In fact, he probably forgot about his margin note altogether!

Whether Fermat had a proof or not, no one else was able to prove the result for over 350 years. And the methods used by Andrew Wiles and company to finally resolve the matter were most certainly not available in Fermat's time. It's also worth noting that no important mathematical issues hinge on the solution of this number-theoretic puzzle. Its importance rests solely on the fame that it attracted after Fermat's death, and on the fact that it remained unsolvable for over three centuries. With this bit of history as background, let's now move to an account of how the Fermat Conjecture actually *became* Fermat's Last Theorem.

INTO THE PAST

When it comes to listing the greatest mathematicians of all time, Leonhard Euler's (1707–1783) name is going to be on that list. No doubt about that. Euler made so many pivotal contributions in so many areas that it's impossible to begin to list them here. It's perhaps some indication of the magnitude and breadth of his work that he is said to have pursued number theory as simply a diversion to his more mainstream work in other areas of mathematics.

Euler received a degree in philosophy at the University of Basel in 1723, and immediately joined the department of theology. But his studies in theology and languages suffered from his devotion to mathematics, and in the autumn of 1726 the Petersburg Academy of Sciences invited him to serve as adjunct of physiology, the only position open at that time. Euler accepted and moved to Petersburg in May of 1727. Despite his having been appointed to study physiology, he was soon given the opportunity to work in pure mathematics. During a 14-year stay in Petersburg, he managed to publish 55 papers in areas as diverse as analysis, number theory, and mechanics.

In 1740, Euler joined the Berlin Academy of Sciences, leaving Petersburg due to the unsettled political situation in Russia at that time. But he returned to Petersburg in 1766, where he remained for the rest of his life. Even though he went blind shortly after his return, Euler was able to continue his work with the aid of assistants. It's of both historical and mathematical interest to look at how Euler proved the case of FLT for the exponent $n = 4$, since it uses basically the same method that Fermat himself used in his own research on the problem, the *method of infinite descent.*

Let's call a positive integer solution (x, y, z) of the equation $x^4 + y^4 = z^2$ an *Euler triple.* Euler wanted to show that no such Euler triples exist. If he could do this, then he would also show that no solution to FLT existed either for $n = 4$, since he could always write z^4 as $(z^2)^2$. To do this, he applied Fermat's principle of infinite descent by defining a procedure that takes as its input any Euler triple, and produces another Euler triple as its output. Moreover, the numbers in the output triple are smaller than those in the input. The existence of such an algorithm for producing successively smaller Euler triples then shows that there are no Euler triples, since that output could then be fed into the algorithm to produce an even smaller Euler triple, and so on. Since this is clearly impossible, as the positive integers have a finite lower bound, no Euler triples exist to get the algorithm going. It's testimony to Euler's genius that he managed to find such an algorithm (which the interested reader can find in the chapter references), thus proving that the equation $x^4 + y^4 = z^4$ has no positive integer solutions. Euler used essentially the same idea to also establish FLT for the case $n = 3$.

The first mathematician to make progress with a *general* approach to FLT was the Frenchwoman, Sophie Germain. Due to the prejudices in the education system against females, Germain was not able to avail herself of the formal study of mathematics. But she managed to use her father's extensive library, sometimes even clandestinely, to educate herself at home. Her best known mathematical work was done in the theory of elasticity, in particular the theory of vibrating surfaces, for which she received a prize of the French Academy of Sciences in 1816. But her major work was in number theory.

Germain outlined her strategy for a general proof of FLT in a long letter to the famed German mathematician, Karl Friedrich Gauss, written on

May 12, 1819. The following idea is central to her approach. Suppose, she argues, that the Fermat equation has a solution (x, y, z) for exponent n. Now suppose that k is a prime number with no nonzero consecutive nth power residues modulo k (that is, the nth powers leave no remainder when divided by k). Then k necessarily divides one of the numbers x, y, or z.

To understand this result, first note that what we mean by an nth power residue modulo k is simply the remainder of a nth power after dividing that power by k. So now suppose the Fermat equation $x^n + y^n = z^n$ has a solution, and suppose that none of x, y, or z is divisible by k. Thus, modulo k, we can divide the entire equation by any of x, y, or z. Letting a represent the multiplicative inverse for x modulo k, we obtain the congruence

$$(ax)^n + (ay)^n \equiv (az)^n \qquad \text{mod } k$$

and thus $1 + (ay)^n \equiv (az)^n$. Consequently, the residues of $(ay)^n$ and $(az)^n$ will be consecutive. Notice that they are also nonzero, since k does not divide a, y, or z. But this contradicts the assumption on k, proving the assertion.

Using this result, Germain concluded that if for a fixed n one could find infinitely many primes k satisfying the condition that k has no nonzero consecutive nth power residues modulo k, then each of these would have to divide one of x, y, or z, and thus one of these three numbers would be divisible by infinitely many primes, which is absurd. This contradiction would then prove FLT for that exponent n.

Despite much effort, Germain was never able to prove FLT by this method *for even a single exponent.* But she did succeed in developing a method for producing many primes k satisfying the above condition. So for any particular exponent n, her method may then show that any solutions to the Fermat equation would have to be quite large. For example, when $n = 5$, she showed that any solutions to the Fermat equation would have to be at least 30 decimals in size.

The only commonly known result of Germain's appeared in 1825, as part of the supplement to the second edition of *Theory of Numbers,* a well-known book on number theory by the famed French mathematician Legendre. To state her theorem would be a bit more technical than we care to be in a book of this type. But the reader can find it in the references cited for this chapter.

Progress on FLT continued very slowly, with proofs sporadically appearing for specific exponents. By the time Peter Dirichlet succeeded in proving the case $n = 14$ in 1832, and the French mathematician Gabriel Lamé for exponent $n = 7$ in 1839, it was clear that their methods would not generalize. A whole new approach was needed.

Such a new method was suggested by Lamé, who announced a proof of FLT at the March 1, 1847 meeting of the Paris Academy of Sciences. Lamé observed that one difficulty encountered in trying to generalize previous approaches was that the second of the factors in the decomposition

$$x^n + y^n = (x + y)(x^{n-1} - x^{n-2}y + x^{n-3}y^2 + \cdots + y^{n-1})$$

has increasingly high degree. He noted that this difficulty could be overcome by factoring the righthand side completely into linear factors, using complex numbers.

Specifically, Lamé used the complex number $r = e^{2\pi i/n}$, the so-called *nth root of unity,* since $r^n = 1$. Using r, we obtain the factorization

$$x^n + y^n = (x + +y)(x + ry)(x + r^2y) \cdots (x + r^{n-1}y).$$

Lamé now wanted to use the same techniques of infinite descent employed by Euler and Fermat to push through a proof. In particular, if (x, y, z) were a solution to the Fermat equation for which the above linear factors were relatively prime, he deduced that since their product is equal to the nth power, z^n, each linear factor is therefore an nth power itself, leading to an infinite descent argument. Such a line of reasoning assumes that if a product of relatively prime numbers is an nth power, then each of the factors is also an nth power. The proof of this requires unique factorization into primes, which is guaranteed for integers by the Fundamental Theorem of Arithmetic. Lamé thought it should hold for complex numbers too. He was wrong.

After some weeks of back-and-forth bickering about Lamé's "proof," the German mathematician Ernst Kummer wrote a letter to the Paris Academy ending the debate. In his letter Kummer showed that Lamé was right—for small values of *n;* unique factorization does hold. But for n greater than 22, it fails—and so, then, does Lamé's proof.

In his study of exactly *why* unique factorization fails for complex numbers, Kummer ended up developing a whole new way of looking at numbers, what is now called the theory of *ideal numbers.* For example, we can factor the number 6 as

$$6 = 2 \times 3 = \left(1 + \sqrt{-5}\right)\left(1 - \sqrt{-5}\right).$$

We might now propose that 2 and $1 + \sqrt{-5}$ have some common, "hidden" divisor. This common divisor would be what Kummer called an "ideal number." Using such ideal numbers, he was able to show that unique factorization failed for large values of the exponent n, thereby negating Lamé's proof of FLT.

Kummer's new theory of numbers did produce positive results on FLT, as well. In fact, he was able to prove FLT for all prime exponents that are *regular.* It's a bit too technical to say here what "regular" means, and it's still unknown whether there are an infinite number of such primes. Nevertheless, for primes less than 100, only 37, 59, and 67 are not regular. Sad to say, though, (for Kummer, anyway), it is known that there are an infinite number of primes that are not regular.

A touch of piquancy was added to FLT in 1908, when details were announced by the Göttingen Royal Society of Science of a prize in the amount of 100,000 gold deutschmarks (around $300,000 in today's terms). This bequest came from the will of Paul Wolfskehl, a banker from Stuttgart. Wolfskehl had studied mathematics under Kummer in Berlin, where he learned of the Fermat conjecture. It's unclear whether Wolfskehl ever tried solving the problem himself, but it's natural to think that this was indeed the case. In any event, it seems that Wolfskehl contracted multiple sclerosis around 1880, and was paralyzed from this illness by 1890. As he was in need of constant care, his friends and family encouraged him to marry an oldish spinster, Susanne Frölich, in 1903. But fate was not on the side of the long-suffering Paul, and his wife made the last years of his life a living hell. It is said that after Wolfskehl's death, his widow lived, rolling in money, together "with an evil maid and an equally evil Doberman" in the Wolfskehl villa in Darmstadt. Perhaps as a kind of small act of revenge for her evil ways, Wolfskehl changed his will in January 1905, bequeathing much of his fortune to "whomsoever first succeeds in proving the Great Theorem of Fermat."

The terms of the Wolfskehl Prize were very specific. For the correct solution of the prize task he laid down the sum of 100,000 marks and decided that the Royal Society of Science in Göttingen should hold in trust this money and serve as judge for awarding of the prize. Wolfskehl explicitly provides for the case that a counterexample to the Fermat conjecture is found. He says that in the case of a counterexample, an acceptable solution must provide a necessary and sufficient condition for those exponents n for which the Fermat equation is unsolvable.

Until Andrew Wiles's solution in 1994, the Göttingen Academy received over 5,000 papers purporting to prove FLT. And each one had to be looked at by a qualified mathematician, and the error in the proposer's argument discovered and pointed out to the proposer. When Andrew Wiles finally received the prize on June 27, 1997, the prize money was 75,000 deutschmarks (about $40,000 at that time).

After Kummer, the next major advance was made in 1922 by British mathematician Leo Mordell, who conjectured that for any integer n greater than 2, there are at most a finite number of solutions to the Fermat equation (including *no* solutions). Basically, what Mordell did was to consider the possible solutions to the Fermat equation in the domain of complex numbers, where it is easier to understand the totality of possible solutions. Initially, evidence for Mordell's Conjecture was pretty scant. So it came as quite a shock to the mathematical community when Gerd Faltings, a young German mathematician now at Princeton, proved the conjecture in 1983.

Shortly after Faltings' remarkable work, which earned him a Fields medal, the mathematical equivalent of the Nobel prize, D. R. Heath-Brown proved that FLT is true for "almost all" powers: if there are any exceptional values of n for which solutions do not exist, then those values must be very thinly distributed as n becomes very large. An excellent illustration of this is the fact that by 1993 FLT had been verified by computer experiments up to values of n larger than 4 million!

The various approaches and attempts to prove FLT led to much good mathematics and in some cases to even great mathematics. But they didn't lead to a solution to the problem. For that much more powerful tools of modern algebra and geometry had to be developed, and assembled in just the right way to crack this particularly resistant nut. The next section outlines how this was done by an international team of mathematicians, culminating in Andrew Wiles's pounding the last nail into the coffin of FLT in 1994. Before detailing Wiles's remarkable achievement, let's look at the background problem that he *really* solved, from which the proof of FLT was a simple corollary.

THE STW CONJECTURE

Some years back, Robert Langlands, a mathematician at the Institute for Advanced Study in Princeton, outlined a program for mathematical research that, if successful, will culminate in a unified theory of a set of objects called zeta functions. These are extremely useful mathematical gadgets having the form

$$\zeta(s) = 1 + 1/2^s + 1/3^s + 1/4^s + \cdots,$$

where s is a complex number having the form $s = a + bi$, a and b are real numbers and i is the square root of -1. This function pops up throughout many branches of mathematics and physics, probably because it is intimately involved with calculation of the distribution of prime numbers. In particular, they are of considerable use in the theory of Diophantine equations in three variables, x, y, and z—which we now know is exactly the class of equations we need to know about to solve the Fermat Conjecture. But until Wiles's work, there was nothing even remotely approximating an overarching theory of Diophantine equations. And this was the fundamental problem Hilbert hoped to correct in his famous Tenth Problem, which we outlined earlier in the opening chapter.

From that earlier discussion, we know that in 1970 Yuri Matyasevich, of the Steklov Mathematical Institute in Leningrad, showed that such a grand theory for all Diophantine equations is impossible. No matter what tricks mathematicians dream up for solving Diophantine equations, there will always remain some equations whose solutions are undecidable. In 1974 Matyasevich and the late Julia Robinson showed that this class of undecidable Diophantine equations includes those with 13 or more variables by producing an example of such an equation for which it is impossible to know if it has a finite or an infinite number of solutions. This number was lowered to nine variables by James Jones of the University of Calgary in 1982. But what about equations with fewer than nine variables? No one knows. The line between solvability and impossibility might be as low as four variables or as high as eight. But mathematicians can now say that the proof of FLT shows that Diophantine equations in three variables should be solvable. In this section we'll see why. That result will then set the stage for the actual proof of FLT by Wiles and Taylor that we will examine in the following section.

At a little-known conference in Japan in 1955, mathematician Yutake Taniyama advanced a conjecture about the equivalence of mathematical objects known as elliptic curves and the mathematics of rigid motions of objects in space. (Note: elliptic curves are *not* ellipses; rather, they are so-called because they are used in calculating the length of arcs along an ellipse, such as the distance the Earth travels in part of its orbit around the Sun). Later, Taniyama's conjecture was sharpened and publicized by Taniyama's friend, mathematician Goro Shimura and famed algebraic geometer André Weil, eventually acquiring the label the *STW Conjecture*, after the surnames of these three mathematicians.

The equivalence asserted by the STW Conjecture is a little tricky to understand. So its useful first to look at a similar connection between two seemingly different ways of looking at a circle. Geometrically speaking, a circle is the set of all points in the (x, y) plane that are equidistant from a single point, the center of the circle. If we plot all such points at a distance 1 from the origin, taken to be the center point, we obtain a curve whose equation is $x^2 + y^2 = 1$. Algebraically, the circle is the set of points (x, y) satisfying this equation.

But there is another way of looking at the circle, a way that associates it with the one-dimensional space of the real number line. Let x be any real number, and consider all numbers of the form $x + nc$, where c is the circumference of the circle and n is any positive or negative whole number. The collection of numbers $\{x + nc\}$ defines an *equivalence class* of numbers, in the sense that each such class defines a set of real numbers that differ from the "base number" x by a fixed distance nc related directly to the circle.

It's certainly not a priori obvious that these two descriptions of a circle—the algebraic formula and a set of equivalence classes of real numbers—have anything to do with each other. But they actually define the *same* set of numbers. Here's why. Suppose f is any periodic function. Then $f(x + cn)$ must have the same value as $f(x)$ for some number c and any integer n. In the case of the circle of radius one, we take f to be either the sine or cosine function. It then follows immediately from the Pythagorean Theorem that $\cos^2 + \sin^2 = 1$. If we then replace $\cos x$ by X and $\sin x$ by Y, we obtain the equation $X^2 + Y^2 = 1$—the original algebraic description of the circle. This means that we have managed to parametrize the equation of the circle by the periodic functions $\cos x$ and $\sin x$.

The STW Conjecture proposes a way of parametrizing not circles, but elliptic curves by a similar type of substitution. The equation for an elliptic curve is $y^2 = x^3 + ax + b$, which is only slightly more complicated than that for the circle. In the nineteenth century, Weierstrass showed how it can be parametrized by generalizing the notion of equivalence classes on the real number line to points in the two-dimensional plane. His method is rather ingenious. Suppose the plane is an infinite sheet of very thin, clear plastic upon which we inscribe the usual horizontal x axis and vertical y axis. Now cover the plane with a grid of regularly spaced parallel lines A units apart in one direction and B units apart in another direction. Note that the lines of this grid do not have to be parallel to the x and y axes. Nor do they have to be at right angles to each other; they need only be parallel. But for simplicity of explanation, let's assume that they are indeed orthogonal. This "tesselation" of the plane then divides it into an infinite number of rectangles of width A and height B.

Now close your eyes and throw a dart at this tesselated plane. Wherever the dart lands, it will be either inside one of the rectangles or on a boundary. Because all rectangles are the same, every one of the infinite number of rectangles must contain exactly one point that corresponds to the place where the dart landed. Thus, any point on the plane can be mapped onto a point in any of the rectangles in the plane. Basically, what this means is that the whole plane can be collapsed into a single rectangle. So the rectangles divide the plane into equivalence classes in exactly the same way the integers divide the real number line (by defining intervals of length one, which serve to define the equivalence classes).

When the plane is squeezed down into a rectangle like this the rectangle acquires some interesting properties. In particular, the parallel sides of the rectangle become equivalent. In other words, when you move from the top to the bottom, you reappear again at the top when you cross the bottom boundary. Similarly, when you move from the left edge of the rectangle to the right edge. This is familiar to most of us from various types of computer games, which identify the top and bottom and left and right edges of the monitor screen in just this fashion. Thus, whereas a circle has a single period, tiling of the plane by rectangles gives rise to two periods—one horizontal, one vertical. To reflect this double periodicity geometrically, we simply cut out the rectangle and then glue its top and bottom edges together. This gives a cylinder, whose two ends are then also glued together. The resulting object is a torus, a doughnut-shaped surface.

The double periodicity of the torus is fairly obvious: the circle that goes around the torus in the "long" direction around the rim, together with the circle that goes around it through the hole in the center. And just as periodic functions can be defined on a circle, doubly periodic functions can be defined on a torus. Weierstrass showed that such doubly periodic functions can be used to parametrize elliptic curves. So by choosing suitable lengths for A and B in the tesselation of the plane, it's possible to restate any elliptic equation in terms of equivalence classes in the plane.

While elegant and easy to visualize, Weierstrass's method is not the only way to parametrize elliptic curves. Shimura, Taniyama, and Weil propose another method for the special class of elliptic curves $y^2 = x^3 + ax + b$, when a and b are integers. The STW Conjecture asserts that, in addition to the torus, there is another surface that can provide the necessary equivalence for this restricted class of elliptic curves. This surface is different for every elliptic curve, but they all resemble a kind of blob poked full of holes, much like a torus with a lot of handles glued onto it. To generate such a surface, it's necessary only to take a polygon (*not* a rectangle) of an appropriate shape, match up pairs of opposite sides, and then glue the sides together in the same way that the torus is formed from a rectangle.

Use of a polygon instead of a rectangle is the key to resolution of the STW Conjecture. As with the rectangle that gives rise to the torus, the polygon provides a method for defining equivalent points in the plane. But now the equivalence classes don't come from a tesselation but from rigid motions of the plane. This is a change that moves the plane without stretching or shrinking or cutting any part of it. For instance, shifting every point in the plane one unit upward is such a motion. So is rotating the plane through the origin by a fixed angle. If you choose one point in the plane and follow its movement in such a motion, there will be exactly one point that it ends up at for each new position of the plane. Thus, an equivalence is set up between the original point and its new location at the end of the rigid motion. Functions that are periodic with respect to rigid motions are called *modular functions (forms).* Shimura, Taniyama, and Weil speculated that by picking the right sequence of modular functions, they could create a surface of points that constitute solutions to any elliptic curve for which a and b are integers. This is completely analogous to the way that picking the right set of trigonometric functions (sine and cosine), it's possible to create a curve—the circle—whose points are solutions to the equation $x^2 + y^2 = 1$.

For the more mathematically inclined, modular forms can be described in the following manner. In mathematics we often encounter certain "special functions" f that obey "functional equations" such as

$$f(xy) = f(x) + f(y)$$

$$f(x + y) = f(x)f(y)$$

$$f(x + 1) = xf(x)$$

Furthermore, if g is any function defined for all complex number $z = x + iy$, then the function

$$f(z) = g(z) + g(1-z) + g(1/z) + g(1/(1-z)) + g(z/(z-1)) + g((z-1)/z)$$

satisfies the functional equation

$$f(z) = f(az + b)/(cz + d))$$

where the quantities a, b, c, and d are allowed to vary over the associated matrix

$$A = \begin{pmatrix} a & b \\ c & d \end{pmatrix}$$

with A being one of the six matrices

$$\begin{pmatrix} 1 & 0 \\ 0 & 1 \end{pmatrix}, \quad \begin{pmatrix} -1 & 1 \\ 0 & 0 \end{pmatrix}, \quad \begin{pmatrix} 1 & 0 \\ 1 & 0 \end{pmatrix},$$

$$\begin{pmatrix} 1 & -1 \\ 1 & 0 \end{pmatrix}, \quad \begin{pmatrix} 0 & 1 \\ -1 & 1 \end{pmatrix}, \quad \begin{pmatrix} 1 & 0 \\ 1 & -1 \end{pmatrix}$$

In the special case when a, b, c, and d satisfy the relation $ad - bd = 1$, the function $f(z)$ is termed "modular."

Regretably, drawing a modular form is impossible. In the case of a rectangular tesselation, we have objects that live in two-dimensional space defined by the x and y axes. A modular form is also defined by two axes—but they are both complex, each having a real and an imaginary part. This means that we need *four* numbers to define each point, not just two, so the modular forms live in a four-dimensional space, what's called *hyperbolic space.*

Hyperbolic space is pretty hard for us three-dimensional beings to really visualize or even comprehend geometrically. So we have to take recourse to algebraic manipulations in order to talk meaningfully about modular forms living in this space. Each such form is built from the same basic ingredients, much like hydrogen, oxygen, and carbon atoms are all built from protons, neutrons, and electrons. It is the proportion of each ingredient that gives the atom its chemical characteristics, and it is the same with modular forms. If we call the basic ingredients of the modular forms $M_1, M_2, M_3, \ldots$ then every modular form can be described by simply listing its modular series, or M-series, which tells how "much" of each ingredient the particular form possesses. A typical M-series would then look like $M_1 = 2, M_2 = 1, M_3 = 3, \ldots$. The amount of each ingredient listed in the M-series is crucial for the property of the particular modular form. Depending on how the amounts change, you can generate an entirely different—but equally symmetric—modular form, or maybe destroy the symmetry altogether and end up with something that's not a modular form at all.

Prior to the recent work by Wiles, Ribet, and others on the STW Conjecture, modular forms stood on their own in the landscape of mathematics. Mostly they were studied because of their symmetry properties, and are of rather recent vintage, having been discovered by Poincaré only near the end of the nineteenth century. In particular, they seemed to have little, if any, connection with the elliptic curves underlying FLT, which date to antiquity. The two objects—elliptic curves and modular forms—lived in completely different areas of mathematics and no one (other than STW) seemed to think they had anything to do with one another. The STW Conjecture claimed that they were not only related, but that they were essentially exactly the *same* objects! Let's now turn to how that equivalence translates into Andrew Wiles's proof of FLT.

THE PATH WITH HEART

In the mid-1980s, Gerhard Frey, a mathematician at the University of Saarbrücken in Germany, showed that any counterexample to FLT—any combination of x, y, and z that would solve the Fermat equation—would make it possible to contruct an elliptic curve that would violate the STW Conjecture *provided* that another, more technical, conjecture were true. In 1986 Kenneth Ribet proved this more technical conjecture, thus establishing that if the STW Conjecture were true, then so was FLT. Ribet's proof dramatically raised the stakes in the centuries-long effort to prove FLT, since while FLT is of little importance by itself, the STW Conjecture, if true, would provide a very powerful tool for number theorists studying the number-theoretic properties of elliptic curves. Finally the link between the STW Conjecture and FLT was in place, setting the stage for Andrew Wiles.

In June 1993, Andrew Wiles arrived at the Newton Institute at the University of Cambridge to speak at a conference on number theory. Rumors of some kind of breakthrough were already abounding even before Wiles's presentation, partly because Wiles, who normally doesn't ask to give lectures at all, had asked to give not just one, but three, hour-long talks. The organizer of the meeting, John Coates, who was also Wiles's doctoral supervisor, scheduled him for Monday-Wednesday, June 21–23.

In his first two lectures, Wiles established the STW Conjecture for a limited subclass of elliptic curves. But to prove FLT he would have to prove the conjecture for all semistable elliptic curves. The excitement

Figure 5.1. Andrew Wiles at the moment of announcing his proof of FLT.

mounted with each passing day, and finally on Wednesday Wiles unveiled a real piece of magic. He showed how to parlay his result for the limited subclass of semistable elliptic curves into a proof of the STW Conjecture for the entire class. Wiles ended his lecture that day by modestly noting that FLT was a simple corollary of his main result. After a moment of stunned silence, the room erupted into applause for this historic moment. Figure 5.1 shows a smiling Wiles at the moment of his historic announcement that day in Cambridge.

As newspapers all over the world trumpeted the demise of this deceptively simple problem, all was not entirely well in the world of Fermat, as a fly in the ointment soon emerged. In early December 1993, Wiles sent out an E-mail confirming that there was a gap in the proof. The gap involved something called an Euler system, which Wiles used to study some crucial properties of the groups associated with the elliptic curves he needed to nail down.

Vague rumors had already been in circulation since early July, just days after Wiles's historic Cambridge lecture, wondering about Wiles's use of Euler systems, though at that time no error had been discovered in the proof. But on November 15, the rumors were acknowledged in a lecture by John Coates, which coincidentally was held in the very lecture room in Cambridge in which Wiles had made his June lectures. Finally, on December 4 Wiles sent out another E-mail acknowledging the gap in the proof. The message stated:

> In view of the speculation on the status of my work on the Taniyama-Shimura conjecture and Fermat's Last Theorem I will give a brief account of the situation. During the review process a number of problems emerged, most of which have been resolved, but one in particular I have not yet settled. The key reduction of (most cases of) the Taniyama-Shimura conjecture to the calculation of the Selmer group is correct. However the final calculation of a precise upper bound for the Selmer group in the semistable case (of the symmetric square representation associated to a modular form) is not yet complete as it stands. I believe that I will be able to finish this in the near future using the ideas explained in my Cambridge lectures.
>
> The fact that a lot of work remains to be done on the manuscript makes it still unsuitable for release as a preprint. In my course in Princeton beginning in February I will give a full account of this work.

Interestingly, the news media did not give the gap in the proof the same kind of attention it gave to Wiles's original announcement. *The New York Times* carried a front page story, complete with a picture of Fermat, the day after Wiles delivered his Cambridge lectures. But it wasn't until a week after Wiles's announcement of the gap in the proof that the *Times* ran a story. And then it was buried on page nine.

In the months after Wiles's Cambridge lectures, the media showered him with attention. He was named one of the "25 Most Intriguing People of 1993" by *People* magazine, along with Princess Diana, President Clinton, and Michael Jackson. Wiles turned down an offer by the Gap, the jeans company, to pose for an advertisement (perhaps in recognition of "the gap" in his proof?). But rumors that movie actress Sharon Stone asked to meet him proved false. Others whose work set the stage for Wiles also had their share of attention. For example, Gerhard Frey was stopped by a U.S. customs official in an airport, who asked, "Are you the same Frey who discovered the connection between Fermat and elliptic curves?" (This sounds even less likely to be true than the rumor about Sharon Stone, at least on the basis of my own dealings with such officials.)

Perhaps the strangest media reaction to Wiles's work came from Marilyn Vos Savant, who is listed in the *Guiness Book of World Records* as having the highest IQ. This woman writes a column for the Sunday newspaper supplement *Parade Magazine.* On November 21, 1993, her column was devoted to showing why Wiles's proof of FLT was incorrect.

Vos Savant points out that the famous problem of "squaring the circle" has been proved to be impossible, so any "proof" of that fact can be assumed to be erroneous. She then steps onto thin ice by implying that János Bolyai's squaring of the circle in hyperbolic geometry was mistaken, because "his hyperbolic proof would not work in Euclidean geometry." Saying that Wiles's proof is "based in hyperbolic geometry," she then applies the same logic to Wiles's work: "If we reject a hyperbolic method of squaring the circle, we should also reject a hyperbolic proof of Fermat's last theorem!!"

How did Vos Savant get the idea that Wiles's proof had anything to do with hyperbolic geometry? Apparently, the connection came via Barry Mazur, a professor at Harvard, and whose work also provided one of the pieces in the Fermat puzzle. Mazur said that following Wiles's Cambridge lectures, the Harvard mathematics department received a request from Vos Savant for information about the proof of Fermat. So he sent her a copy of his paper, "Number Theory as Gadfly," which explains some of the number theory connected to FLT and Wiles's work. Vos Savant then took off and wrote not only her column, but a whole book about FLT published by St. Martin's Press in New York. In the book, Vos Savant thanks not only Mazur, but also Kenneth Ribet and Karl Rubin, even though she never contacted Mazur after receiving the paper, and the other two say they never heard from her at all! Mazur finally wrote a letter to St. Martin's Press denouncing the book and disavowing any involvement with it. Still, the gap in Wiles's proof hung like a sword of Damocles over FLT, the mathematical community wondering whether it would be filled or if FLT would live to fight another day. Enter Richard Taylor.

WHAT THE HECKE

The gap discovered by the experts in Wiles's proof did not invalidate the main part of the work—his proof of the STW Conjecture for large classes of elliptic curves. But it broke the chain of logic for the class of such curves needed to establish FLT.

The problem with the proof lay in the construction of a complicated mathematical object called an Euler system. Such constructions were of recent vintage, having been developed only a few years before by Viktor Kolyvagin in Moscow. In his proof Wiles had to establish a relationship between the sizes of two seemingly unrelated algebraic objects, a "deformation ring" and a "Hecke ring," the latter something associated with modular forms, the former involved with elliptic curves. Wiles had first tried to prove the relationship by a "direct" approach, involving "gluing"

Hecke rings together. But in 1991 he abandoned the direct approach in favor of the more elegant machinery of Euler systems. However the construction based on Euler systems used in Wiles's 1993 result contained a technical flaw—one that could not be fixed. As Wiles put it, "The dragon was showing signs of waking up."

In spring of 1994, Wiles gave a seminar in Princeton in which he outlined his ideas. He also began working at that time with Richard Taylor, his former graduate student, who was then at Cambridge University. They became ever more convinced that the Euler system approach would not work, and decided to return to the "direct" attack. The obstacle that had precipitated his abandonment of this approach—a "local intersection" property for Hecke rings—was still there, however, and by August Wiles and Taylor had come to a dead end. At that point Wiles says he was resigned to a long, arduous journey to the pot of Fermat gold at the end of the FLT rainbow.

In one of those strokes of fate that turn the course of worldly—and mathematical—events, Taylor suggested they take just one more look at the Euler systems. That was the key. Strangely, though, Euler systems themselves *didn't* work. But thinking through exactly *why* they didn't gave Wiles the clue he needed. In Wiles's own words to an audience in Princeton celebrating his victory over "the dragon":

> I was taking one last look at the Euler system, and tried to formulate exactly what was wrong with it. Suddenly, on September 19 last year, I had this wonderful revelation. I saw in a flash how to glue together certain of the Hecke rings. This was the missing key to an old approach to the problem that I'd taken up to 1991, but had then abandoned in favor of Euler systems. My problems were over. I was so amazed by this that for several hours I put it down and did some administrative chore, and then returned to it to check that it was still there. I kept doing this. It was so simple and so elegant that at first it seemed too good to be true. In fact, it was too good to be false.

Wiles's verdict was vindicated by the quick acceptance of the revised paper, prepared together with Taylor, in which the old section on Euler systems was replaced by the simpler Hecke rings approach. The proof now sailed through the review process, as this time the experts found no gaps. The final result, all 130+ pages of it, was published in May 1995 in the *Annals of Mathematics.* As far as the media and the nonmathematical public were concerned, that was the end of the matter. But they were wrong for several reasons.

First of all, the result proved by Wiles and Taylor was *not* FLT. It was a vastly stronger result, from which FLT follows as an incidental corollary. That theorem is much more valuable to mathematics than FLT ever was or could be. For one thing, it marks the first step forward in the Langlands Program. If/when that program is successful, it will end in a unified theory of zeta functions, mathematical objects that pop up in many branches of mathematics and physics. Of more immediate concern is that the Wiles-Taylor theorem may spark off the greatest advance in the history of Diophantine analysis, leading to a general theory of three-variable Diophantine equations. Of course, we know from our earlier remarks and the results of the opening chapter that there can be no completely general theory of such equations. So a theory for the three-variable case would be almost as far as one could ever hope to go. With this tantalizing prospect in mind, let's consider some of the other mathematical directions that Wiles's magnificent achievement has set mathematicians onto.

IT AIN'T OVER TILL IT'S OVER

Dallas banker Andrew Beal is that rarest of all beasts: a banker with a heart. In his case, his heart is in number theory. In 1994, he heard about FLT when he read of its solution by Andrew Wiles. Beal immediately tried his own hand at solving the problem (unsuccessfully!), hoping to find a simpler proof. He then teamed up with mathematician, Daniel Mauldin of the University of North Texas, to hatch the idea of a new prize for a new problem in number theory. What the two of them came up with is a $50,000 bounty for a proof of a more general version of FLT.

The Beal Conjecture. Fermat's original problem states that if two positive integers are raised to the nth power and then added, the sum can never equal an nth power, provided n is greater than 2. But as Beal worked on this problem, he wondered whether a similar statement was true for equations whose exponents vary. For instance, $17^4 + 34^4 = 17^5$, a case when the summands share a common factor and equal a *different* power. Far rarer are cases when the summands are "relatively prime." This means they have no factor in common. Exactly 10 such examples are known:

$$1^n + 2^3 = 3^2 \quad \text{(for any } n\text{)}$$
$$2^5 + 7^2 = 3^4$$
$$7^3 + 13^2 = 2^9$$
$$2^7 + 17^3 = 71^2$$
$$3^5 + 11^4 = 122^2$$
$$17^7 + 76271^3 = 21063928^2$$
$$1414^3 + 2213459^2 = 65^7$$
$$43^8 + 96222^3 = 30042907^2$$
$$33^8 + 1549034^2 = 15613^3$$
$$9262^3 + 15312283^2 = 113^7$$

Note that in all these cases one of the exponents is 2. Mathematicians have found the five "large" solutions only in the last decade. It's a mystery as to why the solutions fall into two groups. As Andrew Granville of the University of Georgia put it, "It's absolutely staggering that there should be five 'big' solutions. Why on Earth is there such a big jump between the sizes of the two sets of solutions?" Why indeed.

Beal has offered $5,000 for a proof that there are no relatively prime solutions in which all the exponents are 3 or greater. And just like the Powerball lottery, the prize will rise by $5,000 each year, up to the maximum of $50,000.

Granville has lectured for several years on the "Beal equation," $a^x + b^y = c^z$. He always writes the 10 solutions on the board, an exercise that always provokes someone in the class to remark, "Have you noticed that, in all the examples, there is an exponent of 2?" Granville claims that the Beal equation has only a finite number of relatively prime solutions for which $1/x + 1/y + 1/z$ is less than 1. Beal's conjecture limits x, y, and z to values greater than 2, Granville allows the smallest exponent to be 2—provided the other two exponents are large enough. Together with Henri Darmon of McGill University in Montreal, he has proved that for any particular x, y, and z that satisfy this condition, there are only a finite number of solutions, and he believes that the same result will hold for all possible sets of exponents. Granville states, "I'd be surprised but not incredibly surprised if [Beal's] conjecture is wrong. But I'd be absolutely devastated if there are infinitely many solutions [that meet his own condition]."

Ronald Graham of AT&T Research says that Wiles's method for proving FLT probably won't work for proving Beal's conjecture. The problem will probably require a brand-new approach that would not only reprove FLT but a whole lot more. Both FLT and the Beal conjecture may turn out to be special cases of a grander hypothesis called the ABC Conjecture. Roughly speaking, this says that whenever two numbers with many repeated factors are added together, the result is a number that has relatively few repeated prime factors and is unlikely to itself be a power. Let's take a longer look at this conjecture to see why it is so important in the mathematical scheme of things.

The ABC Conjecture. The ABC Conjecture deals with pairs of numbers that have no factors in common. That is, they are relatively prime. To state it properly, we need the notion of the *square-free part* of a number n, denoted $sqp(n)$. This is simply the product of all the distinct prime factors of n ignoring those factors that appear more than once. So, for instance, if $n = 18 = 2\times3\times3$, $sqp(18) = 2\times3 = 6$, while if $n = 16 = 2\times2\times2\times2$, the $sqp(16) = 2$. Of course, if n itself is square-free, then $sqp(n) = n$. A particular case of the latter is when n is prime, of course.

Now suppose A and B are two numbers having no common factor, and let C be their sum. Now consider the square-free part of $A \times B \times C$. For example, if $A = 3, B = 7, C = 10$, $sqp(ABC) = 3 \times 7 \times 10 = 210$. Putting numbers in at random for A, B, and C, it usually turns out that $sqp(ABC)$ is greater than C. But not always. For instance, in the case $A = 1, B = 8, C = 9$, we have $sqp(ABC)/C = (1\times2\times3)/9 = 2/3$. David Masser of the University of Basel proved that the ratio $sqp(ABC)/C$ can get arbitrarily small. So if you name any number greater than zero, then there exist integers A, B, and C such that $sqp(ABC)/C$ is smaller than that number. But by changing the situation just slightly, by raising the numerator to the nth power, we get the ABC Conjecture: the ratio $sqp(ABC)^n/C$ reaches a minimum value for any n greater than 1. What's remarkable about the ABC Conjecture is that it offers a way to reformulate an infinite number of Diophantine problems—and solve them. Here's how it can be used to prove FLT.

Suppose FLT is false. In other words, suppose there exist positive integers x, y, z, and k such that $x^k + y^k = z^k$. As we already know, it involves no loss of generality to assume that x^k and y^k have no common factors. Now let's let $A = x^k, B = y^k, C = z^k$, enabling us to write the Fermat equation as $A + B = C$.

According the the ABC Conjecture, for any value of n larger than 1, $sqp(ABC)^n/C$ must be greater than some minimum value. As the conjecture is still unsettled, this minimum value is not known for a given value of n. But as the proof will work whatever this minimum value turns out to be, for the sake of definiteness let's just take $n = 2$ and the associated minimum value to be 1.

Of course, $sqp(ABC)$ is just another way of writing $sqp(x^k y^k z^k)$, which must be less than or equal to the product xyz. But because x and y are less than z, we must have xyz less than z^3. Therefore, $sqp(ABC)$ is less than z^3, and so $sqp(ABC)^2/C$ is less than $(z^3)^2/C$, which is the same as $z^6/z^k = z^{(6-k)}$. But if the ABC Conjecture is true, we can assume $sqp(ABC)^2/C$ is greater than 1, and so $z^{(6-k)}$ is also greater than 1. However, that is a contradiction for any integer k greater than 5. And the only way to remove the contradiction is to change the assumption that FLT is false. So FLT must be true if the ABC Conjecture is true. By going through this same argument again with a smaller value of n, a similar contradiction would be obtained for any integer k greater than 2, thereby proving FLT.

Another direction in which Wiles's work has drawn mathematicians is the more detailed study of elliptic curves. In particular, the methods Wiles introduced open up new possibilities for a general method to determine the number of rational solutions for the cubic equations defining such curves. This, in turn, is intimately related to another famous conjecture in mathematics made by Brian Birch and Peter Swinnerton-Dyer in the 1960s about the behavior of such curves.

The Birch-Swinnerton-Dyer Conjecture. The central challenge of elliptic curves is to find a way to determine all the solutions of the cubic equation defining the curve that are rational. For example, Fermat himself proved that the elliptic equation $y^2 = x^3 - x$ has exactly three solutions: $y = 0$ with $x = 0, 1, -1$. By the same token, the elliptic curve $y^2 = x(x-3)(x+32)$ has many rational solutions (points on the curve). Figure 5.2 shows this curve, together with the line connecting two such points. It can be shown that such a line connecting any two such rational points intersects a third one. This makes it possible to generate all rational points out of just a few of them.

By computing many examples, Birch and Swinnerton-Dyer discovered an interesting relationship between the number of rational points of an elliptic curve and the behavior of an associated analytic function known as an L-function. If the curve had infinitely many solutions, then its associated L-function had the value 0 at a particular point, and conversely. This is their conjecture. This suggested a way to determine whether an

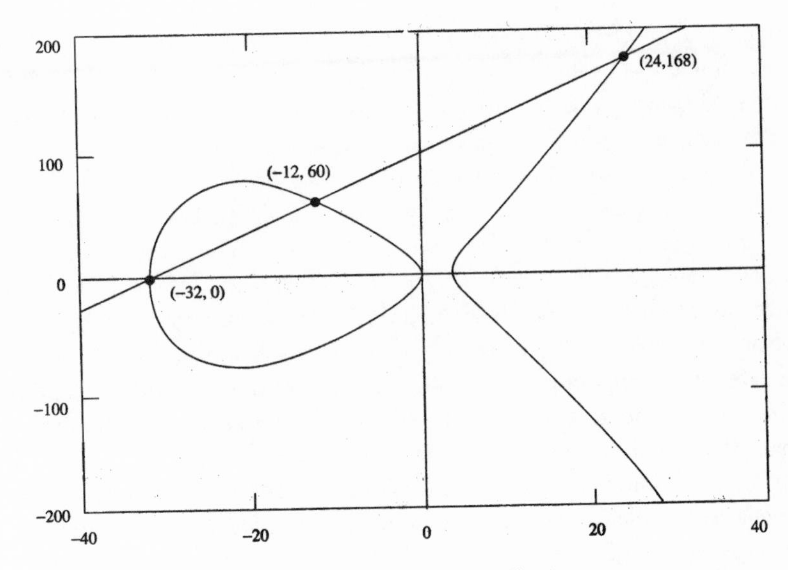

Figure 5.2. Rational points on an elliptic curve.

elliptic curve has an infinite or a finite number of rational points. Just evaluate its L-function at a particular fixed point and see if you get 0.

In 1976, Wiles and John Coates proved half of the conjecture for a special class of elliptic curves—those with a property known as complex multiplication. For this class of curves, they showed that those with an infinite number of rational points always have an L-function that becomes 0 at the critical point. But they were unable to prove the converse, that every curve whose L-function vanishes at that point has an infinite number of rational points.

More recently, Viktor Kolyvagin at the Steklov Institute in Moscow extended the Wiles-Coates theorem to the broader class of modular curves. But the STW Conjecture asserts that *all* elliptic curves are modular. So if it can be established in all its generality, this will establish that every elliptic curve with infinitely many rational points has an L-function that reaches 0 at the appopriate point. But it would still not prove the other direction to draw conclusions about L-functions that vanish. "The other direction is much harder," says Wiles. "My hunch is that using these modular curves is going to be important for completing the proof," he states.

One question that such a proof of the Birch-Swinnerton-Dyer Conjecture would resolve is a puzzle that might have even tempted Pythagoras. It involves the question of whether a given integer can possibly be the area of a right triangle, all of whose sides are of rational length. For example, the integer 6 is the area of the familiar 3–4–5 right triangle. Rather less transparent is that the number 5 is the area of the right triangle whose sides have length 3/2, 20/3, and 41/6. Less obvious—yet still true—is that 1, 2, 3, and 4 are not the areas of any right triangles with rational sides.

The secret of this problem lies in the theory of elliptic curves. Every right triangle with rational sides and area A corresponds to a rational solution of the elliptic equation $y^2 = x^3 - A^2x$. If a, b, and c are the sides and hypotenuse of such a triangle, then some algebraic magic leads to the conclusion that $x = (c/2)^2, y = (a^2 - b^2)c/8$ is a point on the elliptic curve. For the 3–4–5 triangle, the corresponding values of x and y are $x = 25/4, y = 35/8$, which solve the elliptic equation $y^2 = x^3 - 36x$. But there are still no general methods for deciding whether such an equation has rational solutions or not.

The beginnings of a strategy for creating such a general method are suggested by the properties of the equation $y^2 = x^3 - A^2x$, however. First, each such equation has either no rational solutions with nonzero y or infinitely many such solutions. Secondly, these equations have the complex multiplication property, which means the Wiles-Coates Theorem applies to them. Thus, if a curve has an infinite number of rational solutions, its associated L-function must vanish at a special point. So if the L-function does not vanish at this point, the curve has no solutions, implying A is *not* the area of a right triangle with rational sides.

In 1983 Jerry Tunnell of Princeton provided an easy way to do this test, by reducing the evaluation of the L-function to a simple counting problem. But while this method can rule out certain integers as areas of right triangles with rational sides, it can't rule out others. This is because there is, as yet, no proof of the converse of the Wiles-Coates Theorem: it cannot yet be said that every curve whose L-function is 0 at the special point has infinitely many rational solutions. The new tools provided by Wiles's proof of FLT hold the promise of establishing this converse; hence, finally settling the age-old Pythagorean question.

So the longest running, and by far the most famous, unsolved problem in mathematics has finally been laid to rest. *Requiem in pacis, Fermat!*

SUMMARY

The Fermat Conjecture: There are no positive integers x, y, z and $n > 2$ for which the Fermat equation $x^n + y^n = z^n$ has a solution.

Answer: Correct. No such solutions to the Fermat equation exist.

EPILOGUE

THE MAGNIFICENT SEVEN

THE CLAY FOUNDATION PRIZES

On May 24, 2000, at the Collège de France in Paris, the Clay Mathematics Institute attracted worldwide attention by announcing a "bounty" of $1 million dollars each for the solution of seven outstanding mathematical problems. The Institute's aim in establishing these prizes is to celebrate the new millennium and to increase the public visibility of mathematics.

The Clay Mathematics Institute was founded in 1998 by Boston mutual-fund magnate Landon Clay, and is headed by Arthur Jaffe of Harvard University, who is former president of the American Mathematical Society. In remarks at the press conference, Landon Clay tried to explain why he, a businessman with relatively little background in mathematics, is so interested in the field. He cited reasons such as low public funding for mathematics, as well as a personal reason. "Curiosity is part of human nature," he said. "Unfortunately, the established revealed religions no longer provide the answers that are satisfactory, and that translates into a need for certainty and truth. And that is what makes mathematics work, makes people commit their lives to it. It is the desire for truth and the response to the beauty and power and elegance of mathematics that drive mathematicians."

Although based in Cambridge, Massachusetts, the Institute chose Paris as the locale for the announcement of the prize problems in order to celebrate the 100th anniversary of Hilbert's famous lecture at the International Congress of Mathematicians in Paris in 1900, at which he presented his list of 23 Problems important for the development of mathematics in the twentieth century.

But there is an important difference between Hilbert's problems and the Millennium Prize Problems. "Hilbert was trying to guide mathematics by his problems," noted Andrew Wiles of Princeton University, who is a member of the Institute's Scientific Advisory Board. "We're trying to record great unsolved problems. There are big problems in mathematics that are important but where it is very hard to isolate one problem that captures the program," Wiles explained. Jaffe says the Institute is devoted to advancing mathematical knowledge. "If next year every one of these problems were solved, it wouldn't be a problem. But it *would* be a surprise." Let's now have a brief, non-technical look at this list of the problems for mathematical "bounty hunters" in the twenty-first century.

PROBLEMS FOR A NEW CENTURY

P = NP? First on the Clay list is a problem that could make computer encryption a thing of the past. The *P = NP?* problem involves how efficiently computer algorithms process numbers. Consider an algorithm for alphabetizing a list of files. If you double the number of files, the program may then take four times as long, on the average, to put them in order. This is what's called an "n^2 algorithm." Generally, programmers are happy to come up with such "polynomial-time," or *P,* algorithms, as they aren't outrageously time-consuming; the difficulty in solving a problem with such an algorithm goes up only as a polynomial function of the size of the problem's size (measured here by *n,* the number of files in the list).

Even problems that don't seem to be solvable in polynomial time, such as factoring a large number, may be checked in polynomial time. To check whether a number has been factored, all you have to do is multiply the factors together. A problem checkable in polynomial time is called "*NP*" (for "nondeterministic polynomial time). Clearly, all *P* problems are *NP;* if you can solve something in polynomial time, you can certainly check someone else's solution in polynomial time. In 1971, computer scientist Stephen Cook asked whether an *NP* algorithm is necessarily a *P* algorithm.

The answer appears to be "No"; *NP* problems such as factoring large numbers don't have any known polynomial-time solutions. But proving the answer is "No" is another matter. This question acquires even more importance when we realize that mathematicians have proved that the hardest type of *NP* problems, called *NP*-complete, are all equivalent. Thus, a polynomial-time algorithm for one *NP*-complete problem would crack them all—including computer encryptions. Cook says, "You could use one to break any encryption scheme."

The Poincaré Conjecture. In 1904, French mathematician Henri Poincaré was studying the classification of shapes in space. One way to classify these shapes is to look at the behavior of shrinking loops of string on the surface of an object. For example, if you place a loop on the surface of a soccer ball, as it shrinks it will always shrivel up into a point. On the other hand, a loop of string around a doughnut might not be able to shrivel completely; it could get stuck if it is looped through the hole in the doughnut.

For two-dimensional surfaces such as the skin of a soccer ball, the behavior of shrinking loops completely describes the type of surface you're dealing with. If you know that every loop on a given surface shrinks to a point, then the surface is topologically equivalent to a sphere. Poincaré conjectured that the loop-closing test also holds true in the next dimension up, for three-dimensional surfaces. But he never proved or disproved this conjecture—and neither has any other mathematician. Strangely, the conjecture *has* been proven for every other dimension; the three-dimensional case alone remains.

The Birch-Swinnerton-Dyer Conjecture. This problem hinges on the mathematical properties of elliptic curves: the set of points that are solutions of an equation of the form $y^2 = x^3 + ax + b$. The Birch-Swinnerton-Dyer Conjecture, formulated in the 1960s, is concerned with "rational" points on the curve. That is, points where both x and y are rational numbers. Associated with each such elliptic curve is a mathematical object called an L-function, which is a formula that encodes information about the curve in a different form. The conjecture states that there are an infinite number of rational points on a curve if and only if the curve's L-function equals zero at a certain value. Despite the abstract nature of the problem, it is related to questions about the areas of right triangles with sides of rational length, one of the oldest unsolved major problems in mathematics.

The Hodge Conjecture. In algebraic geometry, mathematicians try to combine abstract algebra, which studies the relations and symmetries of numbers (among other things), with geometry, which studies shapes in various spaces. Hodge cycles are structures that have a great deal of algebraic power but no obvious geometric interpretation. They are related to the intersection of curves in space—but are less powerful algebraically. The Hodge Conjecture links the two, stating that a Hodge cycle can be written as the sum of algebraic cycles, combining the power of the former and the easy interpretation of the latter.

Yang-Mills Existence and Mass Gap. This problem is inspired by a branch of physics known as Yang-Mills theory, which describes particles by using the language of mathematical symmetries. Even though Yang-Mills theory has enabled physicists to unify the electromagnetic, weak, and strong forces, it's not certain that reasonable solutions to Yang-Mills equations actually exist. And if they do, whether those solutions will have a "mass gap" that explains why physicists can't isolate quarks.

Navier-Stokes Existence and Smoothness. The Navier-Stokes differential equations describe the motion of incompressible fluids. Although they are relatively simple-looking, the three-dimensional Navier-Stokes equations misbehave very badly. Even with nice, smooth, reasonably harmless initial conditions, the solutions can wind up being extremely unstable. If mathematicians could understand the outrageous behavior of these equations, it would dramatically alter the field of fluid mechanics.

The Riemann Hypothesis. This problem is the granddaddy of all mathematical mysteries. The hypothesis was first published in 1859 by German mathematician Bernhard Riemann, who was investigating the properties of the so-called zeta function: $\zeta(s) = 1 + 1/2^s + 1/3^s + 1/4^s + \ldots$ No matter what positive number you plug in for s, it never makes $\zeta(s)$ equal zero. However, this is not the case if s is a complex number—a number that can be written in the form $s = a + bi$, where i is the square root of -1. In fact, infinitely many zeros all "seem" to have a real part of 1/2; that is, they have the form $1/2 + bi$ for some real number b. The key word here is "seem." Although more than a billion known zeros of the zeta function follow this pattern, no one has proved that they all do.

If the Riemann Hypothesis is true it will affect all of mathematics. For instance, it will tell mathematicians about the distribution of prime numbers. The zeta function is closely related to the L-functions of algebraic geometry, so the Riemann Hypothesis affects the same areas of mathematics as did Andrew Wiles's proof of Fermat's Last Theorem.

THE MILLIONAIRE(S)

Suppose you think you have solved the Riemann Hypothesis or the Yang-Mills problem? How do you get your million dollars from the Clay Institute? First of all, you do *not* submit your proof to the Institute! In fact, the rules for the prizes explicitly state that submissions cannot be made directly to the Institute. To be eligible for a prize, the solution must be published in a recognized journal and have been in print for two years. Following this waiting period, the Institute's Scientific Advisory Board would decide whether to consider the solution for a prize.

The rules for awarding the prizes are flexible enough to handle various situations, such as a simple counterexample that leads to a reformulation of a problem. Generally a prize would not be given for such a counterexample, but the Institute has left itself sufficient discretion to award a prize when a problem is judged to be "well and truly finished," as Andrew Wiles states it.

Some might wonder why the well-known Goldbach's Conjecture is not on the Clay list. Goldbach's Conjecture states that every even number greater than 2 can be written as the sum of two primes. Mostly the reason this problem was not chosen for the Millennium Prize competition is that number theory, the area of mathematics to which Goldbach's Conjecture pertains, is so dominated by the Riemann Hypothesis. But those who think they have a solution to the Goldbach puzzler can take heart. It is the quarry in another million-dollar prize competition, sponsored by the publisher of the novel *Uncle Petros and Goldbach's Conjecture* by Apoltolos Doxiadis. To collect the prize for solving Goldbach's Conjecture, one must submit a published solution to one of the book's publishers, Bloomsbury Publishing Company in the United States and Faber and Faber Limited in the United Kingdom, by March 2002. By way of contrast there is no time limit on the Clay prizes.

SUGGESTED READINGS

THE ART OF THE PROBLEM

Wells, D. "Are These the Most Beautiful?" *Mathematical Intelligencer,* Vol. 12, No. 3, 1990, pp. 37–41. Summary of Wells's questionnaire, as well as comments on the remarks made by the respondents.

Grattan-Guiness, I. "A Sideways Look at Hilbert's Twenty-three Problems of 1900." *Notices of the American Mathematical Society,* Vol. 47, No. 7, August 2000, pp. 752–757.

Halmos, P. "The Heart of Mathematics." *American Mathematical Monthly,* August-September 1980, pp. 519–524.

Stewart, I. "Proof and Beauty." *New Scientist,* 26 June 1999, pp. 29–32.

Epstein, D. and S. Levy. "Experimentation and Proof in Mathematics." *Notices of the American Mathematical Society,* Vol. 42, No. 6, June 1995, pp. 670–674.

HILBERT'S TENTH PROBLEM

Devlin, K. *Mathematics: The New Golden Age.* London: Penguin, 1988. One of the best introductory accounts in print of the spectrum of modern mathematics. The section on the Tenth Problem is excellent, giving lots of motivation and worked examples (some of which are borrowed here!) of Diophantine equations and the tricky way Davis, Robinson, and Matyasevich combined their work to solve the problem.

Matyasevich, Yu. *Hilbert's Tenth Problem.* Cambridge, Mass.: MIT Press, 1993. A definitive account of the Tenth Problem and its solution straight from the horse's mouth, so to speak. Technical in places, as one would expect. But filled with many fascinating examples, side remarks, and history of the problem. A must for anyone interested in this corner of logic and computer science.

Rozenberg, G. and A. Salomaa. *Cornerstones of Undecidability.* Hemel Hempstead, UK: 1994. A very good introduction to undecidability, in general. Focuses mostly on Gödel's results and the Halting Problem of Turing. But there is a very interesting section on the Tenth Problem, including an explicit example of a Diophantine equation associated with the Fermat Conjecture, analogous to the one given in the text for testing prime numbers. Also includes a witty pentalogue at the beginning of the book, which is a fictional conversation between Hilbert, Emil Post, Gödel and two laymen, a traveller and a guide. This sets the stage for the technical stuff that follows.

Mathematical Developments Arising From Hilbert's Problems. Proceedings of Symposia in Pure Mathematics, Vol. 28, Providence, RI: American Mathematical Society, 1976. This is an account of the state-of-play on all 23 of Hilbert's problems as of 1975. It is notable for inclusion of an English translation of Hilbert's famous lecture, as well as the eminence of the scholars who summarize progress on each of Hilbert's "Big Problems." The chapter on the Tenth Problem is written by Martin Davis, Julia Robinson, and Yuri Matyasevich. Need I say more?

Davis, M. "Hilbert's Tenth Problem is Unsolvable." *American Math Monthly,* Vol. 80, No. 3, 1973: 233-269. A smooth, straightforward account of the main results leading up to the solution of the Tenth Problem. Only for mathematicians.

Schöning, U. and R. Pruim. *Gems of Theoretical Computer Science.* Berlin: Springer, 1998. The second chapter of this book is devoted to a step-by-step, *computer-science-based* solution of the Tenth Problem. This shows again that the language of logic and the language of computing machines are completely equivalent. Just two ways of saying the very same thing.

Reid, C. "Being Julia Robinson's Sister." *Notices of the American Mathematical Society,* Vol. 43, No. 12, December 1996: 1486-1492. Very moving and informative account of the life of Julia Robinson, written by her sister, the well-known chronicler of mathematics and mathematicians, Constance Reid.

Chaitin, G. *Algorithmic Information Theory.* Cambridge: Cambridge University Press, 1987. Good account of Chaitin's work on complexity, randomness, and the fantastic number Ω.

Jones, J. et al. "Diophantine Representation of the Set of Prime Numbers." *American Math Monthly*, Vol. 83, No. 6, 1976: 449-464. Source for the amazing Diophantine equation given in the text whose positive values are exactly the set of prime numbers.

Davis, M., H. Putnam, and J. Robinson. "The Decision Problem for Exponential Diophantine Equations." *Annals of Mathematics*, Vol. 74, 1961: 425-436. Source for the breakthrough result leading up to Matyasevich's solution of the Tenth Problem.

Web site: `http://logic.pdmi.ras.ru/Hilbert10`. Web site maintained by the Steklov Mathematical Institute in Saint Petersburg. Contains many original papers about the Tenth Problem, as well as hyperlinks to other source material.

THE FOUR-COLOR THEOREM

Tietze, H. *Famous Problems of Mathematics.* Baltimore, MD: Graylock Press, 1965. A delightful account of many famous problems of mathematics, including the pre-Haken-Appel version of the 4CP. Excellent nontechnical discussion of the 4CP, its history, and various approaches to its solution. Good graphics, too.

Fritsch, R. and G. Fritsch. *The Four-Color Theorem.* New York: Springer, 1998. Detailed mathematical account of the solution to the 4CP. That part is for mathematicians only. But the book's first 40 pages give a first-rate account of the history of the 4CP, along with fairly extensive biographical data on all the important people who have worked on it. This much alone is worth the price of the book.

Devlin, K. *Mathematics: The New Golden Age.* London: Penguin, 1988. One of the best introductory accounts in print of the spectrum of modern mathematics. The section on the 4CP is especially good, giving a detailed, accurate, and easy-to-follow discussion of all the ingredients of the Haken-Appel proof. Very highly recommended—and not just for the 4CP.

Saaty, T. and P. Kainen. *The Four-Color Problem.* New York: Dover, 1986. Easy-to-read, yet technical, discussion of all the ingredients needed to prove the 4CC. Includes much, much more on variations of the problem. Good one-stop reference for just about everything you ever wanted to know about map coloring circa the mid 1980s.

Stemple, J. "The Four-Color Problem," in *Papers in Mathematics,* P. Meyer, ed. *Annals of the New York Acacdemy of Sciences,* Vol. 321, May 1979, pp. 91-101. Quite a good, yet brief, introduction to the ideas of reducibility and discharging as used in the Haken-Appel proof.

Tymoczko, T. "The Four-Color Problem and Its Philosophical Significance." *J. Philosophy,* Vol. 2, February 1979: 57-83. This paper kicked off the debate about mathematical truth being a priori versus a posteriori. The central argument of the paper is that the acceptance of the computer proof forces us to adopt a quasi-empirical account of mathematical practice.

Appel, K. and W. Haken. "The Solution of the Four-Color Map Problem." *Scientific American,* Vol. 237, October 1977: 108-121. Intelligent layman's account of the 4CP and its solution by the solvers.

Appel, K. and W. Haken. *"Every Planar Map is Four Colorable." Contemporary Mathematics,* Vol. 98, Providence, RI: American Mathematical Society, 1989. Full account of the solution for mathematicians. Very illuminating introduction outlining not only the history of the problem, but also the various steps along the way to the final solution.

Seymour, P. "Progress on the Four-Color Theorem," in *Proceedings of the International Congress of Mathematicians,* ICM'94, Zurich, Basel: Birkhäuser, 1995, pp. 183-195. Report of a new proof of the 4CC, as well as progress on some open extensions of the problem. Related results on graph theory are also reported here. For mathematicians only.

MacKenzie, D. "Slaying the Kraken: The Sociohistory of a Mathematical Proof." *Social Studies of Science,* Vol. 29, February 1999: 7-60. Very nice account of the historical and sociological context of the Haken-Appel proof. Gives a very good discussion of the fierce controversy over whether the "proof" should be accepted as a proof. Nice development of the idea of what really constitutes mathematical knowledge.

Cipra, B. "Advances in Map Coloring: Complexity and Simplicity." *SIAM News,* December 1996, 20. Brief introduction to Thomasson's results on map coloring in polynomial time, as well as the list coloring problem.

Swart, E. R. "The Philosophical Implications of the Four-Color Problem." *American Mathematical Monthly,* Vol. 87, (1980): 697-707. Mathematician's retort to the philosophical arguments made by Tymoczko in Reference #6.

THE CONTINUUM HYPOTHESIS

Martin, D. "Hilbert's First Problem: The Continuum Hypothesis." *Proceedings of Symposia in Pure Mathematics,* Vol. 28, Providence, RI: American Mathematical Society, 1978: 81-92. Semitechnical account of the status of the CH as of the late 1970s.

McGough, N. "The Continuum Hypothesis." `www.best.com/ii/math/ch`. Web-based telegraphic summary of the CH. Very readable and notable for its discussion of related topics like the philosophy of the CH and confusions about it in the popular literature.

Dunham, W. *Journey Through Genius.* New York: Wiley, 1990. Chapters 11 and 12 of this award-winning book give a very good introduction to both the CH and to Cantor's work in the transfinite realm. Very readable for nonspecialists.

Rucker, R. *Infinity and the Mind.* London: Paladin Books, 1982. Outstanding discussion of infinity and the transfinite realm of sets and numbers. Also gives the author's account of his meetings with Gödel, and a very good treatment of all of Gödel's work on incompleteness and consistency. Not to be missed.

Wang, H. *Popular Lectures on Mathematical Logic.* New York: Van Nostrand Reinhold, 1981. Intelligent layman's overview of logic, computers, and the various problems and methods connecting the two. Quite readable discussion of the CH, together with an introduction to Cohen's method of "forcing," which established one half of the undecidability of the problem.

Brown, J. *Philosophy of Mathematics.* London: Routledge, 1999. My current favorite book on various philosophies of mathematics. Very well-written account of formalism, platonism, and all the rest, together with many examples that really make these schools of thought come alive. This is popular exposition of the highest order.

Dauben, J. *Georg Cantor: His Mathematics and Philosophy of the Infinite.* Princeton: Princeton University Press, 1979. The definitive biography of Cantor's tormented life and his work on the infinite.

Dauben, J. "Georg Cantor's Creation of Transfinite Set Theory," in *Papers in Mathematics,* Annals of the New York Academy of Sciences, Vol. 321, 1979: 27-44. Compressed account of the full-scale biography cited under the previous reference.

Vilenkin, N. Ya. *In Search of Infinity.* Boston: Birkhäuser, 1995. Extremely entertaining and informative introduction to all aspects of the infinite. Can be read with great profit by anyone with even a high-school knowledge of mathematics.

THE KEPLER CONJECTURE

Hales, T. "Cannonballs and Honeycombs." *Notices of the American Mathematical Society,* Vol. 47, No. 4, April 2000: 440-449. Account of the solution of the Kepler Conjecture straight from the horse's mouth, so to speak. Outstanding exposition, most of which can be easily followed by anyone with even a modest mathematical training. Considers not only the Kepler Conjecture itself, but also the Honeycomb Conjecture and much about foams and the Kelvin Problem.

Hales, T. Web site. `http://www.math.lsa.unich.edu/~hales`. Internet site where all of Hales's papers and discussion of the Kepler Conjecture can be downloaded—including the programs by which the Conjecture was finally resolved. Best one-stop site for information about Kepler's Conjecture and related questions. Of special interest is the paper "An Overview of the Kepler Conjecture," which recounts not only the history of the problem, but also gives a roadmap to the proof constructed by Hales and Ferguson.

Stewart, I. *Problems of Mathematics.* Oxford: Oxford University Press, 1992. Chapter six of this delightful volume gives an excellent overview of the Kepler Conjecture circa 1991. Of course, the problem had not yet been solved as of that time, so Stewart's treatment focuses on the erroneous solution by Hsiang. Nevertheless, the description of the problem, as well as related questions surrounding packing of sand and the shape of snowflakes makes this a "must read" for anyone interested in the Kepler Conjecture.

Peterson, I. "The Honeycomb Conjecture." *Science News,* Vol. 156, No. 4, July 24, 1999. Outstanding layman's exposition of the Honeycomb Conjecture's solution by Hales—and much more. It's hard to believe that so much information about bees, honeycombs, and history can be packed into so little space.

Singh, S. "Packing Them In." *New Scientist,* June 28, 1997: 30-33. Easy-to-read account of the history of the Kepler Conjecture. Unfortunately, the article accepts Hsiang's "proof" and says nothing about the work of

Hales and Ferguson. Of course, the real solution was not announced until two years after publication of this article, so that is not really a mark against it.

Mackenzie, D. "The Proof is in the Packing." *American Scientist*, November-December 1998. Very enlightening expository account of the history of the Kepler Conjecture, Hsiang's false solution, and Hales's resolution of the question.

Johnson, T. "After 300 Years, Computers Facilitate Solution to Kepler Stacking Problem." `http://www.wsws.org/articles/1999/jan-1999/math-j06.shtml`. Web-based paper on Hales's solution. Gives a truly excellent layman's account of the strategy of Hales's proof, along with a discussion of the use of computers in mathmatical argumentation.

Musès, C. "Validating Kepler's Conjecture: A New Approach." *Applied Math & Computation*, Vol. 110 (2000): 99-104. Discusses various classes of proof, arguing that Hales's computer-based proof of the Kepler Conjecture is of the lowest variety, simply a validation. Argues for a proof that illuminates *why* and *how* the result is true, not only that it is so. Provocative.

Lines, M. *Think of a Number.* Bristol, UK: Institute of Physics Publishing, 1993. Chapter 13 of this delightful book gives a fine expository account of the sphere-packing problem. Completely accessible even for high-school students.

FERMAT'S LAST THEOREM

Singh, S. *Fermat's Last Theorem.* London: Fourth Estate, 1997. Runaway best-selling account of Wiles's solution of FLT. Gives a lot of historical background to the Fermat problem. This book betrays its origin as a script for a BBC television production on Wiles's achievement, however, and does very little by way of explaining the basic nature of the proof. So if it's historical background you want, Singh's book is an excellent place to start. But for even a glimpse of the mathematics it leaves much to be desired.

Aczel, A. *Fermat's Last Theorem.* New York: Four Walls Eight Windows, 1996. This little 150-page volume is notable for its painstaking account of questions of priority surrounding the original announcement of the STW Conjecture, including alleged attempts by certain members of the Bourbaki to steal some or all of the credit. As with Singh's book mentioned above, it is long on accounts of Fermat's years on the bench, Galois's fatal duel, Cauchy's legendary arrogance, Poincaré's discovery of automorphic (modular) functions, and other deeds of giants like Turing, von Neumann, and Gödel. And just like Singh, what Aczel softpedals is any coherent account of the structure of the actual proof produced by Wiles and Taylor. Pity. Nevertheless, the book can be recommended to readers who (like the young Andrew Wiles) hunger for these human interest stories surrounding the actual proof. But reading such tales is like eating the frosting on the cake—but forgetting to eat the cake. For that one must turn elsewhere.

Laubenbacher, R. and D. Pengelley. *Mathematical Expeditions.* New York: Springer, 1999. This volume, which is devoted to extracts from original papers, coupled with first-rate exposition and explanation, gives a fantastic account of both the history and solution of FLT in chapter 4. Here in just 47 pages the authors trace the path from the origin of FLT in Fermat's marginal notes to Wiles's solution *without* compromising on the mathematics one bit. Everything you need or want to know about FLT is here. (Incidentally, so is an equally wonderful chapter on the Continuum Hypothesis.)

Cipra, B. *What's Happening in the Mathematical Sciences, 1995–1996,* Volume 3. Providence, RI: American Mathematical Society, 1996. The opening section in this volume is a nice, scientific-journalist account of the final proof of FLT. The writing is clear and the exposition covers all the major steps on the way to Wiles's proof. Notable for a fine explanation of the STW Conjecture and its central role in the proof.

Braner, K. "Paul Wolfskehl and the Wolfskehl Prize." *Notices of the American Mathematical Society,* Vol. 44 (November 1997): 1294-1303. Very nice account of the origin of the Wolfskehl Prize, as well as the thousands of manuscripts that the Göttingen Academy had to review during the course of its existence. This article also contains a lot of interesting sociohistorical information about the practice of mathematics in Europe in the early part of the twentieth century.

Goldfeld, D. "Beyond the Last Theorem." *The Sciences,* March/April 1996: 34-40. First-rate expository account of not only Wiles's solution of FLT, but also both the STW and the ABC Conjectures. Very informative and easy to read.

Rubin, K. and A. Silverberg. "A Report on Wiles' Cambridge Lectures." *Bulletin of the American Mathematical Society,* Vol. 31 (July 1994):

15-38. Detailed, technical account of exactly what Wiles proved and how he did it in his famed 1993 Cambridge lectures. For specialists only!

Faltings, G. "The Proof of Fermat's Last Theorem by R. Taylor and A. Wiles." *Notices of the American Mathematical Society,* Vol. 42 (July 1995): 743-746. Short overview of the final proof of FLT by Wiles and Taylor. Pretty technical—but still readable if you know some mathematics.

THE MAGNIFICENT SEVEN

"Million-Dollar Mathematics Prizes Announced." *Notices of the American Mathematical Society,* Vol. 47, No. 8, September 2000: 877-879.

For more information on the Clay Mathematical Institute and the Millennium Prize problems, see the Institute's web site: `http://www.claymath.org/prize_problems/statement.html`.

Smale, S. "Mathematical Problems for the Next Century." *Mathematical Intelligencer,* Vol. 20, Spring 1998: 7-15. Another excellent set of problems for the new millennium.

INDEX